Crustacean Zooplankton Communities in Chilean Inland Waters

CRUSTACEANA MONOGRAPHS constitutes a series of books on carcinology in its widest sense. Contributions are handled by the Series Editor and may be submitted through the office of KONINKLIJKE BRILL N.V., P.O. Box 9000, NL-2300 PA Leiden, The Netherlands.

Published in this series:

CRM 001 - *Stephan G. Bullard*	Larvae of anomuran and brachyuran crabs of North Carolina
CRM 002 - *Spyros Sfenthourakis et al. (eds.)*	The biology of terrestrial isopods, V
CRM 003 - *Tomislav Karanovic*	Subterranean Copepoda from arid Western Australia
CRM 004 - *Katsushi Sakai*	Callianassoidea of the world (Decapoda, Thalassinidea)
CRM 005 - *Kim Larsen*	Deep-sea Tanaidacea from the Gulf of Mexico
CRM 006 - *Katsushi Sakai*	Upogebiidae of the world (Decapoda, Thalassinidea)
CRM 007 - *Ivana Karanovic*	Candoninae (Ostracoda) from the Pilbara region in Western Australia
CRM 008 - *Frank D. Ferrari & Hans-Uwe Dahms*	Post-embryonic development of the Copepoda
CRM 009 - *Tomislav Karanovic*	Marine interstitial Poecilostomatoida and Cyclopoida (Copepoda) of Australia
CRM 010 - *Carrie E. Schweitzer et al.*	Systematic list of fossil decapod crustacean species
CRM 011 - *Pedro Castro et al. (eds.)*	Studies on Brachyura: a homage to Danièle Guinot
CRM 012 - *Patricio R. De los Ríos-Escalante*	Crustacean zooplankton communities in Chilean inland waters

In preparation (provisional titles):

CRM 013 - *Katsushi Sakai*	Axioidea of the world and a reconsideration of the Callianassoidea (Decapoda, Thalassinidea, Callianassida)
CRM 014 - *Charles H.J.M. Fransen et al. (eds.)*	Studies on Malacostraca: Lipke Bijdeley Holthuis Memorial Volume
CRM 015 - *Akira Asakura et al. (eds.)*	New frontiers in crustacean biology – Proceedings of the TCS Summer Meeting, Tokyo, 20-24 September 2009
CRM 016 - *Danielle Defaye et al. (eds.)*	Studies on freshwater Copepoda: a volume in honour of Bernard Dussart
CRM 00x - *Chang-tai Shih*	Marine Calanoida of the China seas

Author's address:
Prof. Dr. P.R. DE LOS RÍOS-ESCALANTE, Escuela de Ciencias Ambientales, Facultad de Recursos Naturales, Universidad Católica de Temuco, Casilla 15-D, Temuco, Chile; e-mail: patorios@msn.com

Manuscript first received 29 April 2009; final version accepted 7 May 2010.

Cover: Dorsal views of female (left) and male (right) of the large calanoid copepod, *Parabroteas sarsi* (Ekman, 1905) [= *Gigantella sarsi* Ekman, 1905]. [Reproduced from: EKMAN, S., 1905. Cladoceren und Copepoden aus antarktischen und subantarktischen Binnengewässern gesammelt von der Schwedischen Antarktischen Expedition 1901-1903. Wiss. Ergebn. Schwedischen Süd-Polar Exped., **5** (4): 1-40, pls. 1-3; the two figures used here correspond to pl. 2 figs. 13, 20.]

Crustacean Zooplankton Communities in Chilean Inland Waters

By

Patricio R. De los Ríos-Escalante

CRUSTACEANA MONOGRAPHS, 12

BRILL

LEIDEN • BOSTON

This book is printed on acid-free paper.

Library of Congress Cataloging-in-Publication Data

The Library of Congress Cataloging-in-Publication Data is available from the Publisher.

ISBN: 978-90-04-17460-3

PRINTED IN THE NETHERLANDS

CONTENTS

DEDICATORY

Dedicated to my wife, Eliana, for her valuable and infinite support and patience, and to my B.Sc. students, Francisco Correa, Marcela Galindo, Marilyn Gonzalez, Reinaldo Rivera, and Esteban Quinan, who gave me inspiration and support.

PREFACE

The assemblages of crustacean zooplankton in Chilean inland waters are characterized by their low species richness, the presence of both cosmopolitan and endemic species, and a marked predominance of calanoid copepods. The first studies on Chilean planktonic crustaceans were based on scientific expeditions to southern Chilean Patagonia during the last decades of the 19th century and the first decade of the 20th century. In subsequent decades, also the northern Patagonian lakes and the central Chilean lagoons were studied, and only recently the shallow saline and subsaline lagoons in northern Chile have been examined. In spite of these antecedents, there still are zones in which no limnological studies have been performed, e.g., some central Chilean lagoons and wetlands, and the inland waters on many of the oceanic islands that belong to Chile.

In the present book, the current state of our knowledge on the planktonic Crustacea in Chilean inland waters has been gathered from the existing literature, while in addition data from newly explored bodies of water are added in order to present a comprehensive account, as complete as possible, of what is known to date. All species assemblages so far encountered are detailed to their composing faunistic elements, which are invariably treated in an explicitly ecological context.

Temuco, May 2010

PATRICIO R. DE LOS RÍOS-ESCALANTE

INTRODUCTION TO THE PLANKTONIC CRUSTACEA REPORTED FROM CHILEAN INLAND WATERS

Present situation and previous studies

The crustacean zooplankton in Chilean inland waters is characterized by low numbers of species and a marked predominance of calanoid copepods (Soto & Zúñiga, 1991; De los Ríos & Crespo, 2004; De los Ríos & Soto, 2006; Soto & De los Ríos, 2006). This pattern has been observed for northern Chilean lakes (18-27°S; De los Ríos & Crespo, 2004; De los Ríos, 2005), the large, deep lakes in Patagonia (39-51°S; Soto & Zúñiga, 1991; De los Ríos & Soto, 2007), and shallow ponds in central and southern Patagonia (44-54°S; De los Ríos, 2005, 2008; De los Ríos et al., 2008).

The species reported are both cosmopolites and endemics. Among the cosmopolitan or at least widespread species, are the brine shrimp, *Artemia franciscana* (cf. Triantaphyllidis et al., 1998; Gajardo et al., 2004), and cladocerans such as *Daphnia pulex*, *Ceriodaphnia dubia*, and *Eubosmina hagmanni* [= *Neobosmina chilensis*] (cf. Araya & Zúñiga, 1985; Soto & De los Ríos, 2006; De los Ríos & Soto, 2007).

Another remarkable phenomenon is the presence of endemic species, mainly calanoid copepods such as *Tumeodiaptomus diabolicus*, which has been reported from central Chile and northern Patagonia (Araya & Zúñiga, 1985; Schmid-Araya & Zúñiga, 1992; Villalobos et al., 2003; De los Ríos & Soto, 2007). The genus *Boeckella* is present in the area with some widespread species like *B. gracilipes* and *B. michaelseni* (cf. Araya & Zúñiga, 1985; Bayly, 1992a, b; Menu-Marque et al., 2000; De los Ríos & Soto, 2007b), as well as with species restricted to shallow lakes on the Altiplano, such as *B. calcaris*, and species restricted to shallow lakes in southern Patagonia such as *B. brasiliensis* and *B. brevicaudata* (cf. Menu-Marque et al., 2000). A similar situation has been reported for the species of cyclopoid copepods restricted to South America, e.g., *Mesocyclops araucanus* (cf. Gutiérrez-Aguirre et al., 2006).

The history of the studies on crustacean zooplankton in South America started with the first expeditions to southern Patagonia between 1890 and 1910, reported upon in the works of Ekman (1901), Mrázek (1901), and Daday (1902), who made the first descriptions of planktonic crustaceans from the area, confirming or modifying the then existing, sparse records. Those studies formed the basis for much later, in fact quite recent taxonomic investigations (Bayly, 1992a, b; Menu-Marque et al., 2000; Soto & De los Ríos, 2006; Adamowicz et al., 2007) and ecological studies (De los Ríos, 2005, 2008; Soto & De los Ríos, 2006; De los Ríos et al., 2008a, b).

Other sites studied were the large and deep Patagonian lakes and the central Chilean lagoons, and those results were published since 1930 (Brehm, 1935a, b, c, d, 1936, 1937; Löffler, 1962; Pezzani-Hernandez, 1970; Domínguez, 1971, 1973; Zúñiga, 1975; Zúñiga & Domínguez, 1977, 1978; Domínguez & Zúñiga, 1979; Zúñiga & Araya, 1982). These studies were the basis for the compilations of Araya & Zúñiga (1985) and Bayly (1992a, b). Concurrently with those taxonomic descriptions, the first ecological studies were published (Löffler, 1962; Thomasson, 1963), which were carried out mainly in the large, deep Patagonian lakes, and these were, on their turn, a basis for the descriptions of Campos et al. (1982, 1983, 1988, 1989, 1990, 1992a, b, 1994a, b), Campos (1984), Soto & Zúñiga (1991), Soto & Campos (1995), Soto & Stockner (1996), Wölfl (1996), Villalobos et al. (2003), De los Ríos & Soto (2006, 2007a, b), and Soto & De los Ríos (2006).

For central Chilean water bodies, only a few ecological studies have been done, and those mainly in the Rapel reservoir (Zúñiga & Araya, 1983; Ruiz & Bahamonde, 2003), as well as in two small lagoons located close to Valparaíso (Schmid-Araya & Zúñiga, 1992) and Santiago (Mühlhauser & Vila, 1987).

Finally, in northern Chile, with the exception of the descriptions of Brehm (1936), available studies are rather recent and some were done primarily for Bolivian and Peruvian water bodies (Hurlbert & Keith, 1979; Hurlbert et al., 1984, 1986; Zúñiga et al., 1991, 1994; Williams et al., 1995).

Habitats and assemblages of crustacean zooplankton

From a biogeographical point of view, the kind of zooplankton assemblages met with varies along a north-south gradient, as a function of the extended Chilean continental territory (De los Ríos, 2003; Montecino et al., in press), whereas from a hydrological vantage point we may recognize the following zones [adapted from Niemeyer & Cereceda, 1984, and Luebert & Pliscoff, 2006]:

1. Atacama desert. — This zone is located between 18 and 27°S, and it is arid with a few water bodies that are mainly ephemeral streams. In addition, shallow saline lagoons are located mainly between 2000 and 4500 m a.s.l., which are of volcanic origin and associated with saline deposits (Chong, 1988), the brines containing sulphates as their main components (Richaser et al., 1999). These water bodies originally experienced a rather wet climate during the Pleistocene, and the present arid climate developed as a result of weather changes during the last 10 000 years (Grosjean et al., 1996). Currently, the extremely arid conditions characterize this zone as an absolute desert, without vegetation in the coastal zones or in the middle plains, while in the mountain zones there are assemblages of xerophytic shrubs that vary in composition as a function of the local altitude (Luebert & Pliscoff, 2006). Due to the high evaporation, the sparingly present water bodies all show a moderate to high level of salinity (Chong, 1988), and this condition is the main regulating factor for structuring the zooplankton assemblages they harbour (Zúñiga et al., 1991, 1994; De los Ríos & Crespo, 2004; De los Ríos, 2005; De los Ríos & Contreras, 2005).

2. Rivers of a mixed regime in the semi-arid zone. — This zone is located between 27 and 33°S; here we find numerous rivers and approximately six larger reservoirs, mainly at 30-32°S, and small lagoons at 33°S (Niemeyer & Cereceda, 1984). The climate is semi-arid with sporadic rains in winter, and sustains a mixed vegetation of arid and Mediterranean floristic elements. The remaining vegetation mainly comprises crops in agricultural zones in the fluvial valleys and catchment basins (Luebert & Pliscoff, 2006).

3. Rivers of a mixed regime in the sub-humid zone. — This zone is located at 33-37°S and encompasses many rivers and small lagoons, as well as three large reservoirs (Niemeyer & Cereceda, 1984). The prevailing climate is of an oceanic Mediterranean type, with rains increasing particularly in winter in the southern zones. The vegetation in the northern parts is characterized by the presence of shrubs that are gradually replaced in southern direction by *Nothofagus obliqua* (Mirb.) Oerst. forests (Luebert & Pliscoff, 2006).

4. Rivers regulated by lake effluents. — This region is located between 37 and 42°S, and also includes Chiloé Island (Niemeyer & Cereceda, 1984). The area also comprises so-called "Araucanian lakes", which are large, deep, oligotrophic lakes of glacial origin (Campos et al., 1982, 1983, 1987a, b, 1988, 1990, 1992a, b, 1994a, b; Soto et al., 1994; Wölfl, 1996; Villalobos, 1999), and from these lakes the rivers of this zone originate. The water bodies of the region include those in the Nahuelbuta mountains that have small oligotrophic and mesotrophic lakes located at 38°S (Muñoz et al., 2001), the coastal wetlands of the Araucania region (Hauenstein et al., 2002), and a number of small, pristine mountain lakes (Steinhart et al., 2002). The climate is temperate oceanic, with *Nothofagus obliqua* forests, while in the mountain zones between 37 and 39°S there are *Araucaria araucana* (Molina) K. Koch forests, and south of 40°S, mainly in the mountain zones, forests of *Fitzroya cupressoides* (Molina) Johnston are found (Luebert & Pliscoff, 2006). The large lakes of this region have agricultural zones in their surroundings, and for that reason some of these lakes show a transition from oligotrophy to mesotrophy (Wölfl et al., 2003), whereas in zones with native forest the lakes are pristine and unpolluted (Steinhart et al., 1999, 2002; De los Ríos et al., 2007).

5. Northern Patagonian trans-Andean rivers. — This zone is located between 42 and 48°S, and has oligotrophic, large, and deep lakes (Niemeyer & Cereceda, 1984; Soto & Zúñiga, 1991) as well as small, oligotrophic lakes and shallow ponds (De los Ríos, 2008). Some of the large lakes are shared with Argentina, and from those lakes many large rivers originate that give rise to lakes contained both in their westerly and easterly effluents (Niemeyer & Cereceda, 1984). The climate is again temperate oceanic, with *Nothofagus antarctica* (G. Forster) Oerst., *N. betuloides* (Mirb.) Oerst., and *N. pumilio* (Poeppig & Endler) Krasser forests (Luebert & Pliscoff, 2006).

6. Ice fields of northern Patagonia. — This zone is located between 48 and 53°S. It is characterized by the presence of two main ice fields, with associated lakes and rivers (Niemeyer & Cereceda, 1984). Perhaps one of the most famous sites of this region is the Torres del Paine National Park, which is located in a transition zone at the border of the southern ice field. Here we find a native forest of *Nothofagus antarctica* and semi-arid plains with *Festuca* (cf. Luebert & Pliscoff, 2006). This park has large oligotrophic lakes of glacial origin, small mesotrophic lakes, and shallow temporal, ephemeral ponds (Soto et al., 1994; Soto & De los Ríos, 2006), as well as peatlands (Henríquez, 2004). The climate is semi-arid, with shrubs in the southern zones. There are rains in winter, and in spring and summer the area experiences strong winds (Soto et al., 1994; Luebert & Pliscoff, 2006). This last condition, obviously greatly enhancing evaporation, would be one of the causes of the presence of ephemeral ponds, and also of more permanent sub-saline as well as saline ponds (Soto et al., 1994).

7. Tierra del Fuego island and the southern islands. — This zone is located to the south of the Strait of Magellan (Niemeyer & Cereceda, 1984). The climate is characterized by rains in winter, the vegetation is formed by

Festuca on the semi-arid plains in the north, while the southern zone has native perennial forests of *Nothofagus antarctica* (cf. Luebert & Pliscoff, 2006). On the island of Tierra del Fuego, there are lakes and many small, shallow lagoons (De los Ríos et al., 2008). Also here, saline lakes may well be present due to the increase of evaporation as a result of the exposition of the island to the strong winds (De los Ríos, 2005).

A CHECKLIST OF THE BRANCHIOPODA (ANOSTRACA AND CLADOCERA)

Chapter summary. — Existing reports on branchiopods from Chilean inland waters are restricted to studies on Anostraca and Cladocera, whereas unfortunately there are no published studies on Notostraca or Conchostraca. The present chapter provides a checklist of the species of anostracans and cladocerans reported from the area. Only two genera of Anostraca have been described from Chile, i.e., the halophilic *Artemia*, which is widespread across the Chilean territory, and *Branchinecta* that has been described from some restricted zones, inhabiting shallow ponds in both northern Chile and southern Patagonia. The cladocerans, are, until now, known to be present with five families encompassing 28 species, distributed over a wide array of various kinds of inland water bodies. These species include cosmopolitic and endemic species, some of the widespread being *Ceriodaphnia dubia* and *Daphnia pulex*, while among the endemics are *Daphniopsis chilensis*, described only from northern Chilean water bodies, and *Daphnia dadayana*, occurring in shallow ponds in southern Patagonia. The present study describes the geographical distribution of Cladocera in Chilean inland waters as recorded to date, and points out the need for more extensive taxonomic and biogeographical studies on the Chilean Branchiopoda.

Introduction

The Branchiopoda of Chilean inland waters have been relatively poorly studied. The only branchiopod taxa studied are anostracans (Soto, 1990; Gajardo et al., 1995; De los Ríos, 2005; De los Ríos et al., 2008; Rogers et al., 2008) and cladocerans (Araya & Zúñiga, 1985; Ruiz & Bahamonde, 1989; Villalobos, 1994). The Anostraca mainly inhabit shallow ponds such as inland saline water bodies (*Artemia*; Gajardo et al., 1998; De los Ríos & Zúñiga, 2000), while species of the genus *Branchinecta* inhabit shallow ephemeral pools (Rogers et al., 2008), characterized by low conductivity and by oligotrophy (De los Ríos et al., 2008a). The cladocerans, in contrast, inhabit a wide range of water bodies such as large, deep lakes (Villalobos, 1997, 2003;

De los Ríos & Soto, 2006, 2007a, b) and shallow ponds, both permanent and ephemeral (De los Ríos, 2005, 2008; Soto & De los Ríos, 2006).

List of Branchiopoda reported from Chilean inland waters

Class BRANCHIOPODA Latreille, 1817
- Order ANOSTRACA G. O. Sars, 1867
  - Family ARTEMIIDAE Grochowski, 1896
    - Genus *Artemia* Leach, 1819
      - *Artemia franciscana* Kellogg, 1906
      - *Artemia persimilis* Piccinelli & Prosdocimi, 1968
  - Family BRANCHINECTIDAE Daday, 1910
    - Genus *Branchinecta* Verrill, 1869
      - *Branchinecta gaini* Daday, 1910
      - *Branchinecta granulosa* Daday, 1902
      - *Branchinecta palustris* Birabén, 1946
      - *Branchinecta papillata* Rogers, De los Ríos & Zúñiga, 2008
      - *Branchinecta valchetana* Cohen, 1981
      - *Branchinecta vuriloche* Cohen, 1985
- Order CLADOCERA Latreille, 1829
  - Family SIDIDAE Baird, 1850
    - Genus *Diaphanosoma* Fischer, 1850
      - *Diaphanosoma chilense* Daday, 1902
    - Genus *Latonopsis* G. O. Sars, 1888
      - *Latonopsis occidentalis* Birge, 1891
  - Family DAPHNIIDAE Straus, 1828
    - Genus *Ceriodaphnia* Dana, 1853
      - *Ceriodaphnia dubia* Richard, 1894
    - Genus *Daphnia* O. F. Müller, 1785
      - *Daphnia ambigua* Scourfield, 1947
      - *Daphnia dadayana* Paggi, 1999
      - *Daphnia obtusa* Kurz, 1874
      - *Daphnia peruviana* Harding, 1955
      - *Daphnia pulex* Leydig, 1860
    - Genus *Daphniopsis* G. O. Sars, 1903
      - *Daphniopsis chilensis* Hann, 1986
    - Genus *Moina* Baird, 1850
      - *Moina micrura* Kurz, 1874
    - Genus *Scapholeberis* Schoedler, 1858
      - *Scapholeberis spinifera* (Nicolet, 1849)
    - Genus *Simosa* (Norman, 1903)
      - *Simosa exspinosa* (De Geer, 1778)
      - *Simosa serrulata* (Koch, 1841)
      - *Simosa vetula* (O. F. Müller, 1776)
  - Family BOSMINIDAE Baird, 1845
    - Genus *Eubosmina* Seligo, 1900
      - *Eubosmina hagmanni* (Stingelin, 1904) [= *Neobosmina chilensis* (Daday, 1902)]

Family MACROTHRICIDAE Norman & Brady, 1867
Genus *Ilyocryptus* G. O. Sars, 1862
Ilyocryptus spinifer Herrick, 1884
Genus *Macrothrix* Baird, 1843
Macrothrix hirsuticornis Norman & Brady, 1867
Macrothrix inflata Daday, 1902
Macrothrix laticornis (Jurine, 1820)
Macrothrix odontocephala Daday, 1902
Macrothrix palearis Harding, 1955
Family CHYDORIDAE Stebbing, 1902
Genus *Alona* Baird, 1850
Alona affinis Leydig, 1860
Alona cambouei De Guerne & Richard, 1893
Alona guttata G. O. Sars, 1862
Alona intermedia G. O. Sars, 1862
Alona poppei Richard, 1897
Alona pulchella King, 1853
Alona quadrangularis (O. F. Müller, 1776)
Genus *Alonella* G. O. Sars, 1862
Alonella clathratula (G. O. Sars, 1896)
Alonella excisa (Fischer, 1854)
Genus *Camptocercus* Baird, 1843
Camptocercus rectirostris (Schoedler, 1862)
Genus *Chydorus* Leach, 1816
Chydorus sphaericus (O. F. Müller, 1785)
Genus *Leydigia* Kurtz, 1874
Leydigia leydigi (Schoedler, 1863)
Genus *Paralona* Šrámek-Hušek, Straškraba & Brték, 1962
Paralona nigra (G. O. Sars, 1862)
Genus *Pleuroxus* Baird, 1843
Pleuroxus aduncus (Jurine, 1820)
Pleuroxus scopulifer Ekman, 1900

Records of species reported

ANOSTRACA

Family ARTEMIIDAE

Artemia franciscana Kellogg, 1906 (fig. 1): Salar de Surire (18°48′S 69°04′W), Salar de Llamara (21°18′S 69°37′W), Yape pools (20°40′S 70°15′W), Salar de Atacama: Cejas lagoon (23°02′S 68°13′W), Tebenquiche lagoon (23°07′S 68°16′W), Chaxas pools (23°09′S 68°13′W), Pampilla pools (29°50′S 71°22′W), Palo Colorado (Los Vilos) pools (31°51′S 71°25′W) (Zúñiga et al., 1999); El Convento salt works (33°52′S 71°44′W) (De los Ríos & Zúñiga, 2000); Pichilemu salt works (Gajardo et al., 1995); La Rinconada lagoon (Crespo & De los Ríos, 2004).

Artemia persimilis Piccinelli & Prosdocimi, 1968 (fig. 1): Amarga lagoon (50°29′S 72°45′W) (Gajardo et al., 1998); De los Cisnes lagoon (53°14′S 70°00′W) (De los Ríos, 2005).

Family BRANCHINECTIDAE

Branchinecta gaini Daday, 1902 (fig. 1): Kon Aikén ephemeral pools, I (52°50′S 71°10′W) (Rogers et al., 2008); Kon Aikén ephemeral pools, II (52°51′S 70°55′W) (De los Ríos et al., 2008).

Branchinecta granulosa Daday, 1902 (fig. 1): Vega del Toro ephemeral pools (51°07′S 71°40′W) (Rogers et al., 2008).

Branchinecta palustris Birabén, 1943 (fig. 2): Salar de Coposa (21°17′S 68°53′W) (Rogers et al., 2008).

Branchinecta papillata Rogers, De los Ríos & Zúñiga, 2008 (fig. 2): pools close to Salar de Coposa (20°40′S 68°42′W) (Rogers et al., 2008).

Branchinecta valchetana Cohen, 1981 (fig. 2): Paniri volcano (22°08′S 68°15′W) (Rogers et al., 2008).

Branchinecta vuriloche Cohen, 1985 (fig. 2): Balmaceda pools (45°53′S 71°40′W) (Rogers et al., 2008).

CLADOCERA

Family SIDIDAE

Diaphanosoma chilense Daday, 1902 (fig. 3): Chungará (18°14′S 69°09′W) (Paggi, 1978); Catapilco reservoir (32°38′S 71°27′W), Peñuelas lagoon (33°09′S 71°32′W), Orozco reservoir (33°14′S 71°25′W), Plateado reservoir (33°04′S 71°39′W), Rapel reservoir (34°10′S 71°29′W), Lanalhue lake (37°55′S 73°19′W), Caburgua lake (39°07′S 71°47′W), Villarrica lake (39°16′S 72°07′W), Panguipulli lake (39°41′S 72°15′W), Pirihueico lake (39°56′S 71°48′W), Ranco lake (40°12′S 72°22′W), Puyehue lake (40°39′S 72°30′W), Rupanco lake (40°49′S 72°30′W) (Araya & Zúñiga, 1985); Lagunitas lagoon (41°28′S 72°56′W) (De los Ríos, 2003); Chapo lake (41°27′S 72°30′W) (Villalobos et al., 2003).

Latonopsis occidentalis Birge, 1891 (= *Latonopsis australis* G. O. Sars, 1888) (fig. 3): El Peral lagoon (33°30′S 71°35′W) (Araya & Zúñiga, 1985).

Family DAPHNIIDAE

Ceriodaphnia dubia Richard, 1894 (fig. 3): Pocuro (32°53′S 70°38′W), Pichilafquén lagoon (39°13′S 72°07′W), Quillelhue lagoon (39°33′S 71°32′W), Villarrica lake (39°16′S 72°07′W), Puyehue lake (40°39′S 72°30′W) (Löffler, 1962); Yeso reservoir (33°39′S 70°07′W) (Pezzani-Hernández, 1970); Plateado reservoir (33°04′S 71°39′W) (Domínguez & Zúñiga, 1976); Rapel reservoir (34°10′S 71°29′W) (Zúñiga & Araya, 1982); Peñuelas lagoon (33°09′S 71°32′W), Orozco reservoir (33°14′S 71°25′W), El Peral lagoon (33°30′S 71°35′W), Negra lagoon (33°39′S 70°08′W), Aculeo lagoon (33°50′S 70°55′W), Lanalhue lake (37°55′S 73°19′W), Lleulleu lake (38°08′S 73°19′W), Caburgua lake (39°07′S 71°47′W), Calafquén lake (39°31′S 72°08′W), Panguipulli lake (39°41′S 72°15′W), Neltume lake (39°47′S 71°59′W), Pirihueico lake (39°56′S 71°48′W), Rupanco lake (40°49′S 72°30′W), Atravezado lake (45°45′S 72°54′W), Lynch lagoon (53°58′S 69°27′W) (Araya & Zúñiga, 1985); Paso (51°02′S 72°55′W), Redonda (51°02′S 72°55′W), Larga (51°02′S 72°55′W) (Soto & De los Ríos, 2006).

Daphnia ambigua Scourfield, 1947 (fig. 3): Plateado reservoir (33°04′S 71°39′W), Peñuelas lagoon (33°09′S 71°32′W), Orozco reservoir (33°14′S 71°25′W), Yeso reservoir (33°39′S

70°07′W), Negra lagoon (33°39′S 70°08′W), Lanalhue lake (37°55′S 73°19′W), Caburgua lake (39°07′S 71°47′W), Villarrica lake (39°31′S 72°08′W), Pellaifa lake (39°36′S 71°57′W), Panguipulli lake (39°41′S 72°15′W), Neltume lake (39°47′S 71°59′W), Riñihue lake (39°49′S 72°19′W), Pirihueico lake (39°56′S 71°48′W), Bonita lagoon (40°53′S 72°54′W), Ranco lake (40°12′S 72°22′W), Puyehue lake (40°39′S 72°30′W) (Araya & Zúñiga, 1985).

Daphnia dadayana Paggi, 1999 [= *D. sarsi* Ekman, 1902; *D. commutata* Ekman, 1900] (fig. 4): Chiguay lagoon (45°56′S 71°50′W) (Villalobos, 1994); Balmaceda pools (45°53′S 71°40′W), Isidoro lagoon (50°57′S 72°53′W), Don Alvaro lagoon (51°01′S 72°52′W), Monserrat lagoon (51°07′S 72°47′W), Vega del Toro pools (51°07′S 72°40′W), Kon Aikén pools (52°50′S 71°10′W): Porvenir pool (52°50′S 70°10′W) (De los Ríos, 2005); Juncos (51°01′S 72°52′W), Jovito (51°02′S 72°54′W), Paso (51°02′S 72°55′W), Redonda (51°02′S 72°55′W), Larga (51°02′S 72°55′W), Cisnes (51°02′S 72°55′W) (Soto & De los Ríos, 2006).

Daphnia obtusa Kurz, 1874 (fig. 4): Del Inca lagoon (32°49′S 70°09′W); Tierra del Fuego island (note details in Villalobos, 1994).

Daphnia peruviana Harding, 1955 (fig. 4): Parinacota (Korinek & Villalobos, 2003).

Daphnia pulex (Leydig, 1860) (fig. 4): Chungará (18°14′S 69°09′W) (Domínguez, 1973); Negra lagoon (33°39′S 70°08′W), Riñihue lake (39°49′S 72°19′W), Pirihueico lake (39°56′S 71°48′W), Ranco lake (40°12′S 72°22′W), Rupanco lake (40°49′S 72°30′W), Todos los Santos lake (41°06′S 72°15′W), Polux lake (45°43′S 71°53′W) (Araya & Zúñiga, 1985); Llanquihue lake (41°07′S 72°50′W) (Löffler, 1962).

Daphniopsis chilensis Hann, 1986 (fig. 5): Pond at Licancabur volcano (Hann, 1986).

Moina micrura Kurz, 1874 (fig. 5): Chungará (18°14′S 69°09′W) (Domínguez, 1973); (Domínguez, 1971); Rapel reservoir (34°10′S 71°29′W) (Zúñiga & Araya, 1982); Rungue reservoir (33°01′S 70°54′W), Peñuelas lagoon (33°09′S 71°32′W), Orozco reservoir (33°14′S 71°25′W), Yeso reservoir (33°39′S 70°07′W), Lanalhue lake (37°55′S 73°19′W) (Araya & Zúñiga, 1985).

Scapholeberis spinifera (Nicolet, 1849) [= *Daphnia spinifera* Nicolet, 1849] (fig. 5): Pocuro (32°53′S 70°38′W), Ranco lake (40°12′S 72°22′W), Calafquén lake (39°31′S 72°08′W), Panguipulli lake (39°41′S 72°15′W), Pellaifa lake (39°30′S 71°57′W) (Löffler, 1962); Riñihue lake (39°49′S 72°19′W) (Campos et al., 1974), Caburgua lake (39°07′S 71°47′W), Pichilafquén lagoon (39°13′S 72°12′W), Neltume lake (39°47′S 71°59′W) (Araya & Zúñiga, 1985); Los Patos pond (39°10′S 71°42′W) (De los Ríos et al., 2007).

Simosa exspinosa (De Geer, 1778) [= *Daphnia exspinosa* (De Geer, 1778)] (fig. 5): Riñihue lake (39°49′S 72°19′W) (Campos et al., 1974); Peñuelas lagoon (33°09′S 71°32′W), El Peral (33°30′S 71°35′W) (Araya & Zúñiga, 1985).

Simosa serrulata (Koch, 1841) (fig. 6): El Peral (33°30′S 71°35′W) (Araya & Zúñiga, 1985).

Simosa vetula (O. F. Müller, 1776) [= *Simocephalus vetulus* (O. F. Müller, 1776); *Daphne vetula* O. F. Müller, 1776] (fig. 6): Chungará (18°14′S 69°09′W) (Domínguez, 1973); Chungará (18°14′S 69°09′W) (Araya & Zúñiga, 1985).

Family BOSMINIDAE

Eubosmina hagmanni (Stingelin, 1904) [= *Neobosmina chilensis* (Daday, 1902); *Bosmina coregoni chilensis* Daday, 1902] (fig. 6): Pocuro (32°53′S 70°38′W), Pichilafquén lagoon (39°13′S 72°12′W), Villarrica lake (39°16′S 72°07′W) (Daday, 1902); Calafquén lake (39°31′S 72°08′W), Pellaifa lake (39°30′S 71°57′W), Riñihue lake (39°49′S 72°19′W), Ranco lake (40°12′S 72°22′W), Puyehue lake (40°39′S 72°30′W), Bonita lagoon (40°53′S 72°52′W),

Llanquihue lake (41°07′S 72°50′W), Todos los Santos lake (41°46′S 73°15′W) (Löffler, 1962); Chungará lake (18°14′S 69°09′W) (Domínguez, 1973); Plateado reservoir (33°04′S 71°39′W) (Domínguez & Zúñiga, 1976); Rapel reservoir (34°10′S 71°29′W) (Zúñiga & Araya, 1982); Catapilco reservoir (32°38′S 71°27′W), Rungue reservoir (33°01′S 70°54′W), Peñuelas lagoon (33°09′S 71°32′W), Orozco reservoir (33°14′S 71°25′W), Yeso reservoir (33°39′S 70°07′W), Negra lagoon (33°39′S 70°08′W), Aculeo lagoon (33°50′S 70°55′W), Lanalhue lake (37°55′S 73°19′W), Lleulleu lake (38°08′S 73°19′W), Caburgua lake (39°07′S 71°47′W), Huilipilún lake (39°08′S 72°10′W), Calafquén lake (39°31′S 72°08′W), Panguipulli lake (39°41′S 72°15′W), Neltume lake (39°47′S 71°59′W), Pirihueico lake (39°56′S 71°48′W), Atravezado lake (45°45′S 72°54′W), La Paloma lake (45°55′S 72°15′W) (Araya & Zúñiga, 1985); Chapo lake (41°27′S 72°30′W) (Villalobos et al., 2003); Elizalde lake (45°46′S 72°55′W), General Carrera lake (45°50′S 72°00′W) (De los Ríos & Soto, 2007); Del Toro lake (51°12′S 72°38′W) (Campos et al., 1994); Juncos (51°01′S 72°52′W), Jovito (51°02′S 72°54′W), Paso (51°02′S 72°55′W), Redonda (51°02′S 72°55′W), Larga (51°02′S 72°55′W), Cisnes (51°02′S 72°55′W) (Soto & De los Ríos, 2006); Isidoro lagoon (50°57′S 72°53′W), Guanaco lagoon (51°01′S 72°50′W), Monserrat lagoon (51°07′S 72°47′W), Vega del Toro pools (51°07′S 72°40′W), Kon Aikén pools (52°50′S 71°10′W); Porvenir pool (52°50′S 70°10′W) (De los Ríos, 2005).

Family MACROTHRICIDAE

Ilyocryptus spinifer Herrick, 1884 (fig. 6): Isla Margarita (41°46′S 73°15′W), Bonita lagoon (40°53′S 72°52′W), Llanquihue lake (41°07′S 72°50′W) (Zúñiga & Domínguez, 1977); Peñuelas lagoon (33°09′S 71°32′W), Orozco reservoir (33°14′S 71°25′W) (Araya & Zúñiga, 1985).

Macrothrix hirsuticornis Norman & Brady, 1867 (fig. 7): Manzanillo river (33°26′S 70°28′W) (Brehm, 1936); Plateado reservoir (33°04′S 71°39′W), Yeso reservoir (33°39′S 70°07′W) (Domínguez & Zúñiga, 1976).

Macrothrix inflata Daday, 1902 (fig. 7): Peñuelas lagoon (33°09′S 71°32′W) (Araya & Zúñiga, 1985).

Macrothrix laticornis (Jurine, 1820) (fig. 7): El Peral (33°30′S 71°35′W) (Araya & Zúñiga, 1985).

Macrothrix odontocephala Daday, 1902 (fig. 7): Del Inca lagoon (32°49′S 70°09′W) (Araya & Zúñiga, 1985).

Macrothrix palearis Harding, 1955 (= *Echinisca palearis*) (fig. 8): Chungará lake (18°14′S 69°09′W) (Domínguez, 1973).

Family CHYDORIDAE

Alona affinis Leydig, 1860 (fig. 11): Peñuelas lagoon (33°09′S 71°32′W), Plateado reservoir (33°04′S 71°39′W), Rapel reservoir (34°10′S 71°29′W) (Araya & Zúñiga, 1985).

Alona cambouei De Guerne & Richard, 1893 (fig. 8): Chungará lake (18°14′S 69°09′W) (Domínguez, 1973).

Alona guttata G. O. Sars, 1862 (fig. 8): Peñuelas lagoon (33°09′S 71°32′W) (Araya & Zúñiga, 1985).

Alona intermedia (G. O. Sars, 1862) (fig. 8): Picton island (Brehm, 1962).

Alona poppei (Richard, 1897) (fig. 9): Villarrica lake (39°16′S 72°07′W) (Araya & Zúñiga, 1985).

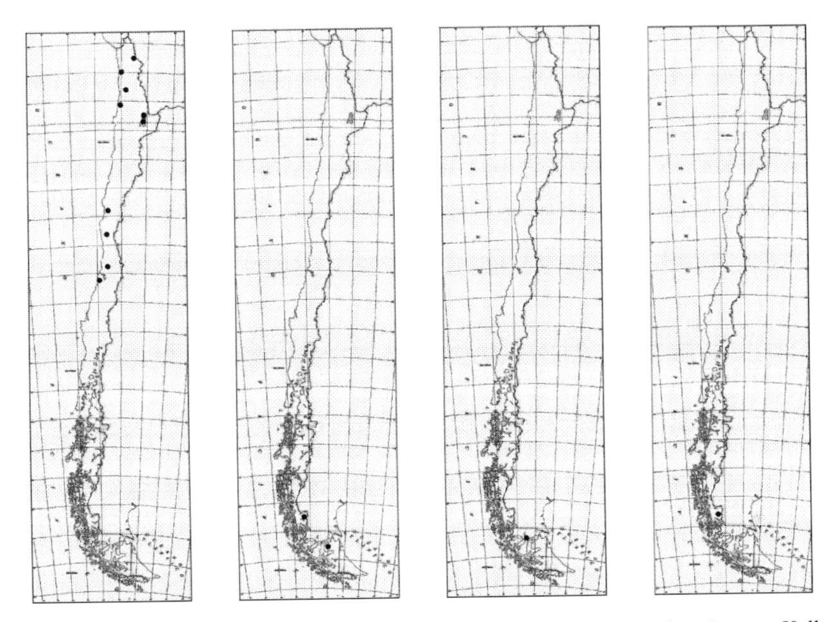

Fig. 1. Records of the following species (left to right): *Artemia franciscana* Kellogg, 1906; *Artemia persimilis* Piccinelli & Prosdocimi, 1968; *Branchinecta gaini* Daday, 1910; *Branchinecta granulosa* Daday, 1902.

Alona pulchella (King, 1953) (fig. 9): Villarrica lake (39°16′S 72°07′W), Pellaifa lake (39°30′S 71°57′W), Llanquihue lake (41°07′S 72°50′W), Todos los Santos lake (41°46′S 73°15′W) (Araya & Zúñiga, 1985).

Alona quadrangularis (O. F. Müller, 1776) (fig. 9): Puyehue lake (40°39′S 72°30′W) (Löffler, 1962).

Alonella clathratula (G. O. Sars, 1896) (fig. 9): Chungará lake (18°14′S 69°09′W) (Domínguez, 1973); Llanquihue lake (41°07′S 72°50′W) (Löffler, 1962).

Alonella excisa (Fisher, 1854) (fig. 10): Todos los Santos lake (41°46′S 73°15′W) (Löffler, 1962).

Camptocercus rectirostris (Schoedler, 1862) (fig. 10): Margarita island (41°06′S 72°17′W), Llanquihue lake (41°07′S 72°50′W), Pellaifa lake (39°30′S 71°57′W) (Löffler, 1962); Del Inca lagoon (32°49′S 70°09′W), Pichilafquén lagoon (39°13′S 72°12′W) (Araya & Zúñiga, 1985); Plateado reservoir (33°04′S 71°39′W), Orozco reservoir (33°14′S 71°25′W), El Peral (33°30′S 71°35′W), Rapel reservoir (34°10′S 71°29′W), Lanalhue lake (37°55′S 73°19′W) (Araya & Zúñiga, 1985).

Chydorus sphaericus (O. F. Müller, 1785) [= *Lynceus sphaericus* O. F. Müller, 1785] (fig. 11): Aculeo lagoon (33°50′S 70°55′W), Ranco lake (40°12′S 72°22′W) (Brehm, 1936); Pichilafquén lagoon (39°13′S 72°12′W), Calafquén lake (39°31′S 72°08′W), Llanquihue lake (41°07′S 72°50′W) (Löffler, 1962); Yeso reservoir (33°39′S 70°07′W) (Pezzani-Hernández, 1970); Chungará lake (18°14′S 69°09′W) (Domínguez, 1973); Plateado reservoir (33°04′S 71°39′W) (Domínguez & Zúñiga, 1976); Riñihue lake (39°49′S 72°19′W) (Zúñiga & Domínguez, 1978); Rapel reservoir (34°10′S 71°29′W) (Zúñiga & Araya, 1982); Peñuelas lagoon (33°09′S 71°32′W), El Peral lagoon (33°30′S 71°35′W), Negra lagoon (33°39′S 70°08′W) (Araya & Zúñiga, 1985); Del Inca lagoon (32°49′S 70°09′W) (Löffler, 1962).

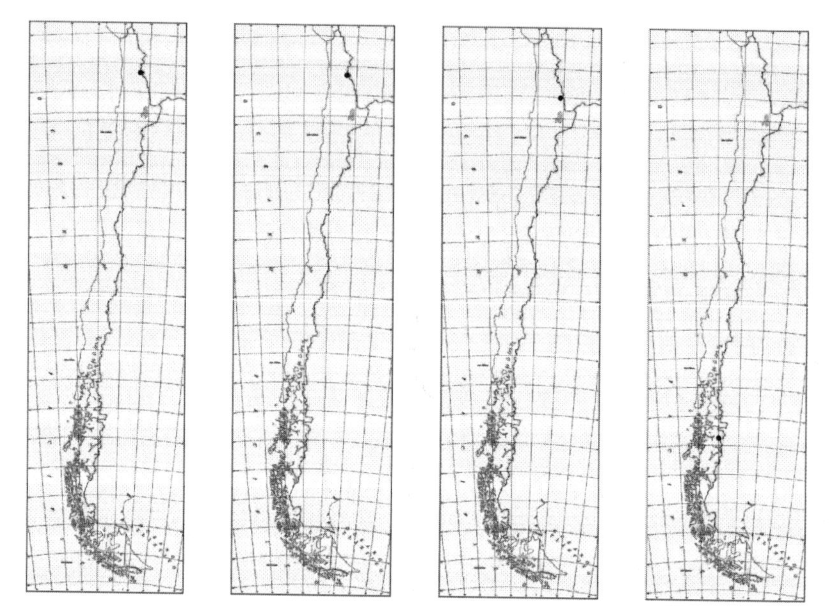

Fig. 2. Records of the following species (left to right): *Branchinecta palustris* Birabén, 1946; *Branchinecta papillata* Rogers, De los Ríos & Zúñiga, 2008; *Branchinecta valchetana* Cohen, 1981; *Branchinecta vuriloche* Cohen, 1985.

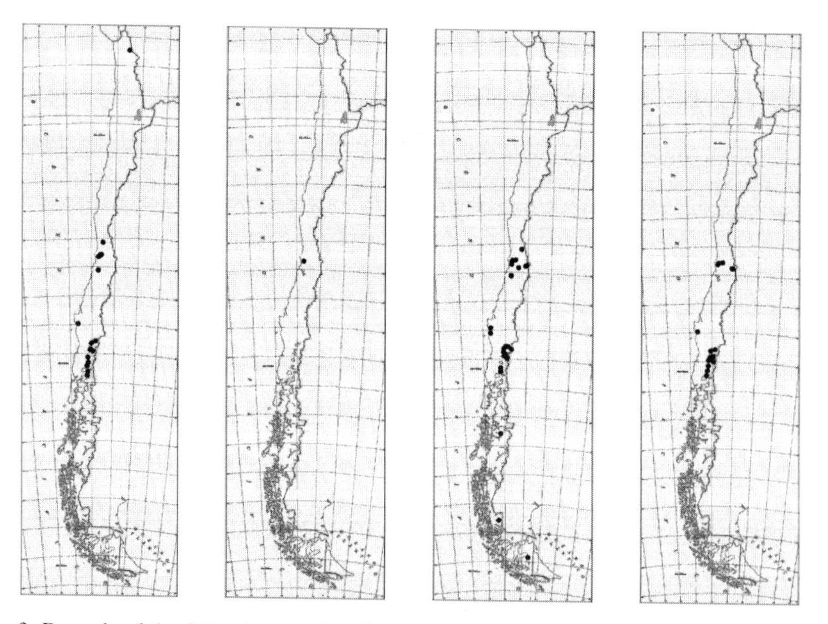

Fig. 3. Records of the following species (left to right): *Diaphanosoma chilense* Daday, 1902; *Latonopsis occidentalis* Birge, 1891, *Ceriodaphnia dubia* Richard, 1894, *Daphnia ambigua* Scourfield, 1947.

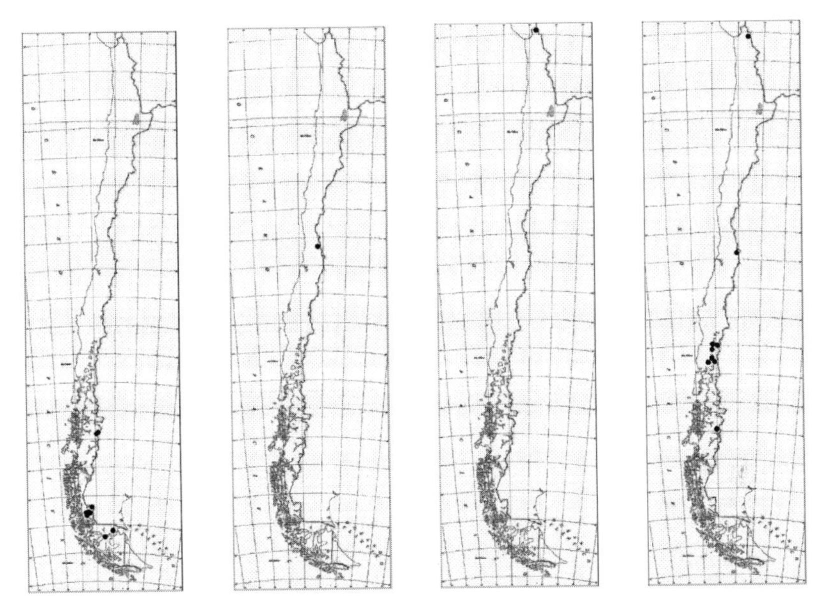

Fig. 4. Records of the following species (left to right): *Daphnia dadayana* Paggi, 1999; *Daphnia obtusa* Kurz, 1894; *Daphnia peruviana* Harding, 1955; *Daphnia pulex* Leydig, 1860.

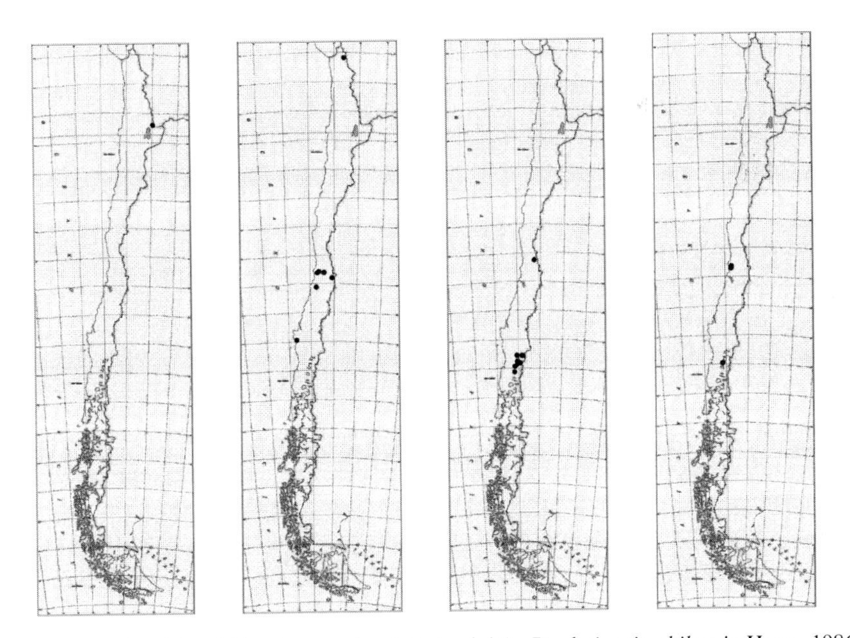

Fig. 5. Records of the following species (left to right): *Daphniopsis chilensis* Hann, 1986; *Moina micrura* Kurz, 1874; *Scapholeberis spinifera* (Nicolet, 1849); *Simosa exspinosa* (De Geer, 1778).

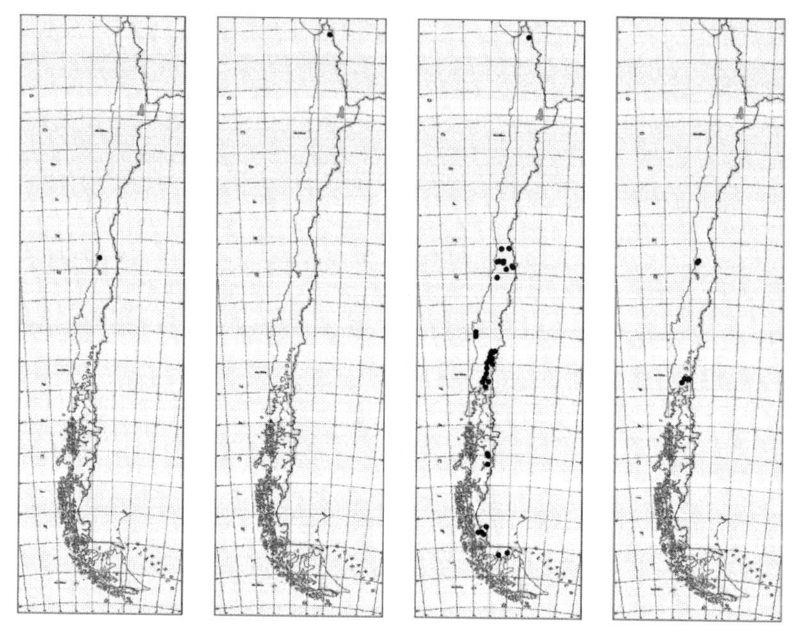

Fig. 6. Records of the following species (left to right): *Simosa serrulata* (Koch, 1841); *Simosa vetula* (O. F. Müller, 1776); *Eubosmina hagmanni* (Stingelin, 1904) [= *Neobosmina chilensis* (Daday, 1902)]; *Ilyocryptus spinifer* Herrick, 1884.

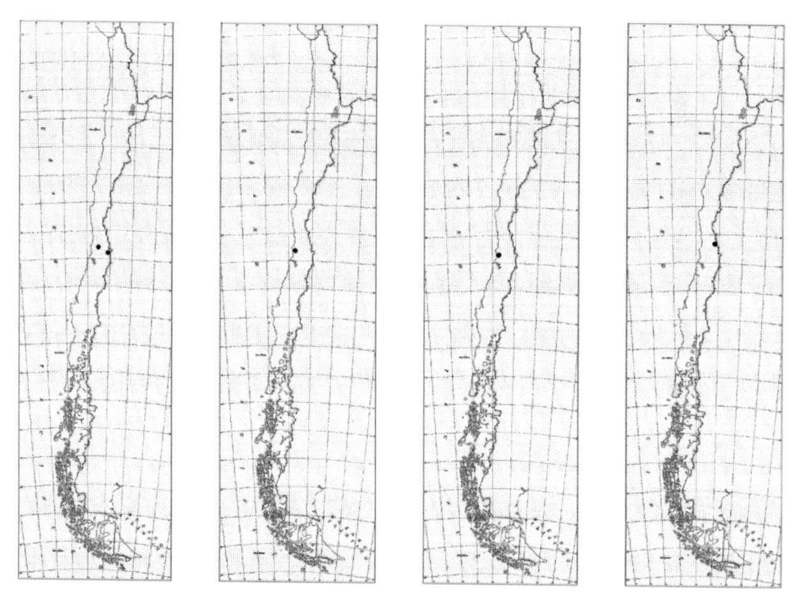

Fig. 7. Records of the following species (left to right): *Macrothrix hirsuticornis* Norman & Brady, 1867; *Macrothrix inflata* Daday, 1902; *Macrothrix laticornis* (Jurine, 1820); *Macrothrix odontocephala* Daday, 1902.

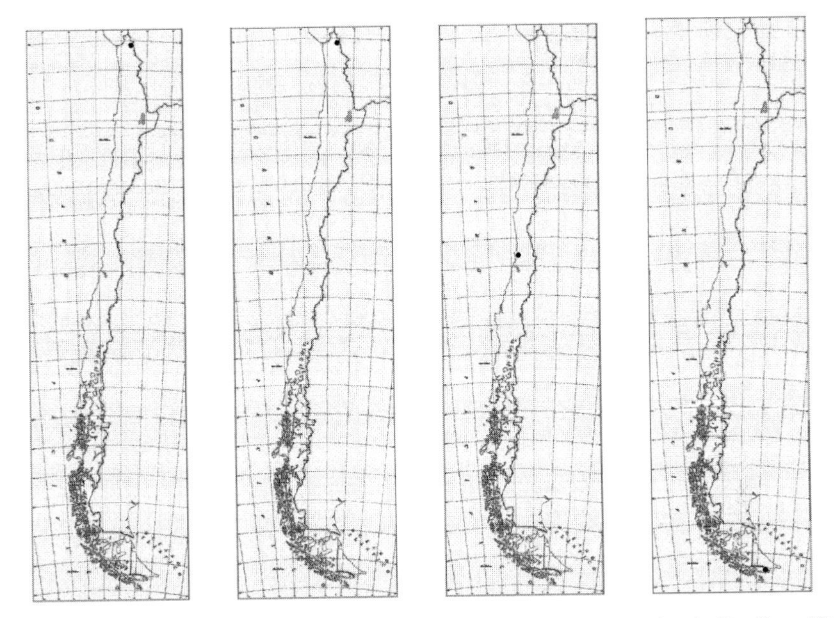

Fig. 8. Records of the following species (left to right): *Macrothrix palearis* Harding, 1955; *Alona cambouei* De Guerne & Richard, 1893; *Alona guttata* G. O. Sars, 1862; *Alona intermedia* G. O. Sars, 1862.

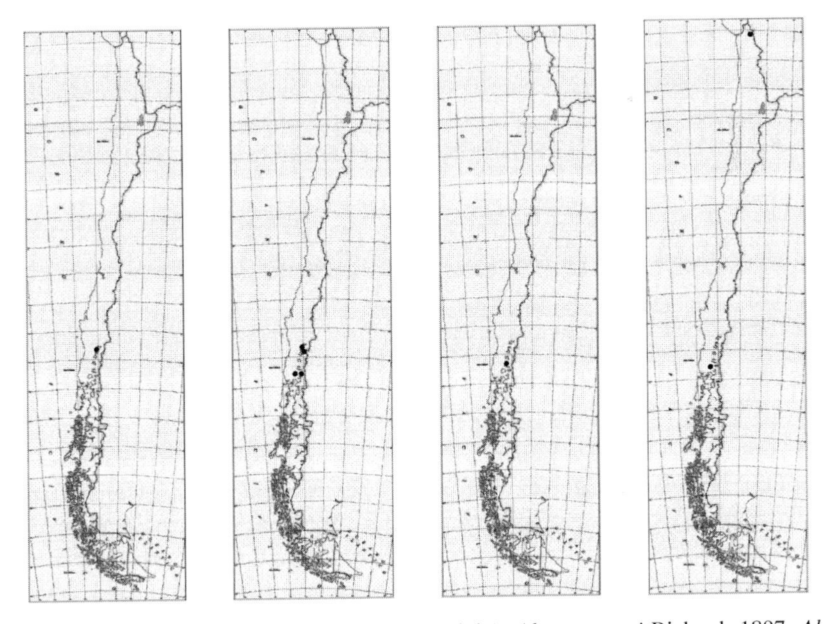

Fig. 9. Records of the following species (left to right): *Alona poppei* Richard, 1897; *Alona pulchella* King, 1853; *Alona quadrangularis* (O. F. Müller, 1776); *Alonella clathratula* (G. O. Sars, 1896).

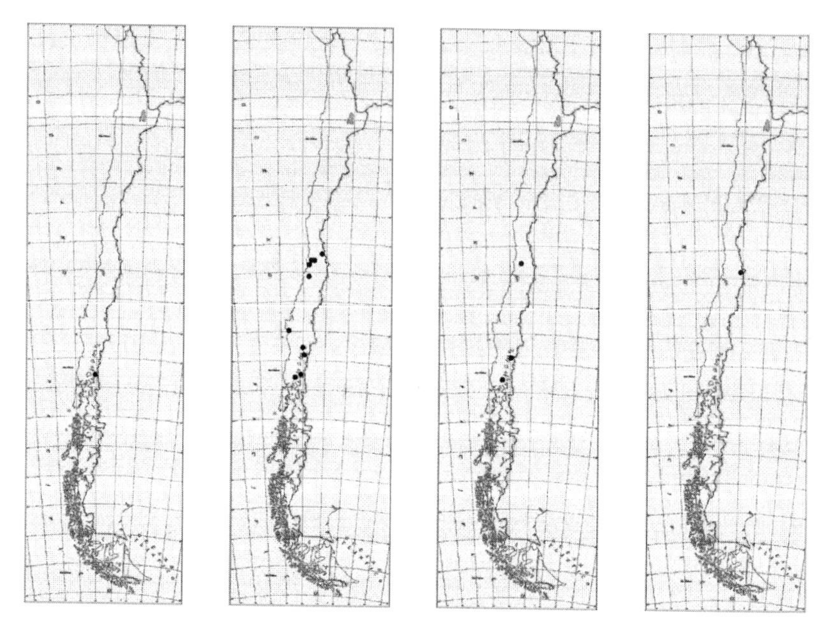

Fig. 10. Records of the following species (left to right): *Alonella excisa* (Fischer, 1854); *Camptocercus rectirostris* (Schoedler, 1862); *Pleuroxus aduncus* (Jurine, 1820); *Pleuroxus scopulifer* Ekman, 1900.

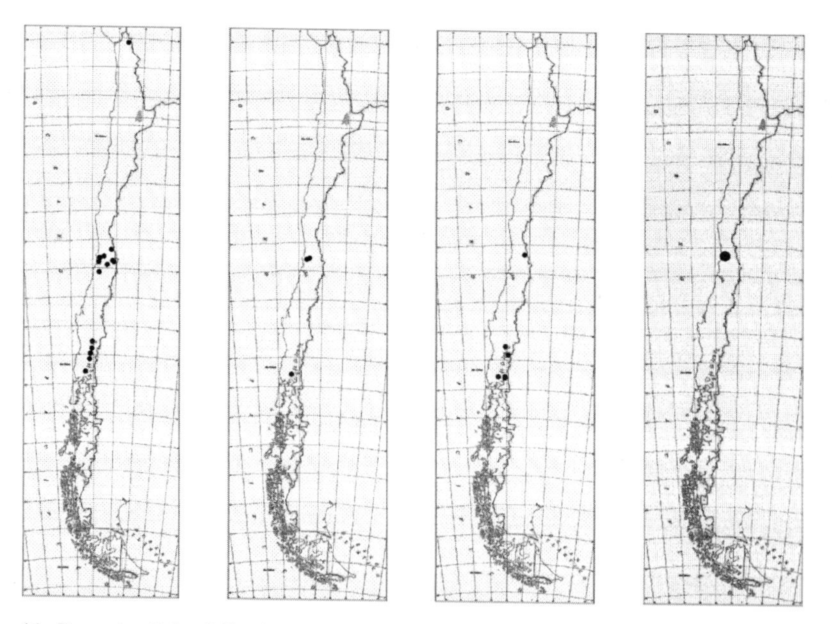

Fig. 11. Records of the following species (left to right): *Chydorus sphaericus* (O. F. Müller, 1785), *Leydigia leydigi* (Schoedler, 1863), *Paralona nigra* (G. O. Sars, 1862), *Alona affinis* Leydig, 1860.

Leydigia leydigi (Schoedler, 1963) (fig. 11): Plateado reservoir (33°04′S 71°39′W) (Domínguez & Zúñiga, 1976); Peñuelas lagoon (33°09′S 71°32′W), Llanquihue lake (41°07′S 72°50′W) (Araya & Zúñiga, 1985).

Paralona nigra (G. O. Sars, 1862) (fig. 11): Del Inca lagoon (32°49′S 70°09′W), Pichilafquén lagoon (39°13′S 72°12′W), Margarita island (41°06′S 72°17′W), Llanquihue lake (41°07′S 72°50′W) (Löffler, 1962); Pellaifa lake (39°30′S 71°57′W), Todos los Santos lake (41°46′S 73°15′W) (Zúñiga & Domínguez, 1977).

Pleuroxus aduncus (Jurine, 1820) (fig. 10): Plateado reservoir (33°04′S 71°39′W) (Domínguez & Zúñiga, 1976) (Löffler, 1962); Pellaifa lake (39°30′S 71°57′W) (Zúñiga & Domínguez, 1977); Llanquihue lake (41°07′S 72°50′W) (Thomasson, 1963, fide Ruiz & Bahamonde, 1989).

Pleuroxus scopulifer (Ekman, 1900) (fig. 10): Yeso reservoir (33°39′S 70°07′W) (Pezzani-Hernández, 1970).

The Anostraca reported from Chile

Artemia franciscana was the first species of its family described from Chilean saline inland waters (Gajardo et al., 1992), and for many years it was supposed to be the only species of *Artemia* present in Chile (Gajardo et al., 1995). However, this species does only occur north of 34°S (Gajardo et al., 2004). It is reported from shallow, saline-sulphate ponds in the Andes mountains of northern Chile between 18 and 23°S, from shallow coastal ponds at 18-31°S, and from artificial salt works in central Chile (33-34°S), the last mentioned habitat probably concerning intentional inoculations (Gajardo et al., 1995, 2004). The other species of *Artemia*, *A. persimilis*, was originally described from Argentina and Italy (Triantaphyllidis et al., 1998), but its main populations are located in southern Argentina, and during many years the literature mentioned this species as endemic for Argentina (Triantaphyllidis et al., 1998). Nevertheless, this species was reported from southern Chilean saline lakes in the Magallanes region (51-54°S; De los Ríos, 2005), specifically in Amarga Lagoon, a saline lake located in the Torres del Paine National Park (Campos et al., 1996; Gajardo et al., 1998, 2000; De los Ríos & Zúñiga, 2000) and in the De los Cisnes lagoon, a shallow saline lake located close to the town of Porvenir, on Tierra del Fuego island (Medina & Báez, 1998; De los Ríos, 2005).

As regards the fairy shrimps, the studies about the species of this order for Chilean inland waters are only scarce. The first reference found merely reported the presence of anostracans in shallow ephemeral ponds in the Magallanes region (Soto, 1990). Specialized literature on this group only mentions the presence of the genus *Branchinecta* for southern South America

(Brték & Mura, 2000). Zooplankton accounts of De los Ríos (2005) and De los Ríos & Rivera (2007) confirm the existence of this genus for shallow ephemeral ponds in the Magellan region, further detailed in a complete taxonomic identification by Rogers et al. (2008). Also, Rogers et al. (2008) reported the first records of fairy shrimps for northern Chilean shallow ponds. About the ecology of *Branchinecta* there is only very limited information, in fact restricted to southern Chilean Patagonia. These references indicate that the genus would occur only in shallow, ephemeral water bodies, oligotrophic and with low conductivity values (De los Ríos, 2005, 2008; De los Ríos et al., 2008a, b).

The Cladocera reported from Chile

The species reported from the Daphniidae exhibit the presence of cosmopolitan species that are also widespread in Chilean inland waters, such as *Daphnia ambigua*, *D. obtusa*, and *D. pulex* (cf. Araya & Zúñiga, 1985; Villalobos, 1994; De los Ríos & Soto, 2007), and some species restricted to distinct zones, like northern Chile, e.g., *Daphniopsis chilensis* (cf. Hann, 1986), and *Daphnia dadayana*, which occurs in shallow lagoons in southern Patagonia (De los Ríos, 2005, 2008). The results presented also reveal the occurrence of a considerable number of species at 33°S. For Bosminidae, which would have one species inhabiting Chilean inland waters (Deevey & Deevey, 1970), the currently compiled references indicate a total of four species (at least, nominal species) to describe the bosminid populations encountered in Chilean inland waters (Soto & Zúñiga, 1991; Campos et al., 1982, 1983, 1987a, b, 1988, 1990, 1992a, b, 1994a, b; Wölfl, 1996; De los Ríos & Soto, 2007b). For Camptocercidae there are only reports of the occurrence of species from that family, but unfortunately no ecological or other data are available. Finally, the family Chydoridae is represented here by some widespread species such as *Alona guttata*, *A. pulchella*, and *Chydorus sphaericus*.

From a biogeographical view point, there are three endemic species in northern Chile (*Daphnia peruviana*, *Daphniopsis chilensis*, and *Macrothrix palearis*), whereas there is a considerable number of widespread species. This northern part of Chile would have a low species richness of Cladocera probably as a result of the high to at least moderate salinity levels measured in the local inland water systems (De los Ríos & Crespo, 2004; De los Ríos, 2005). On the other hand, many species have been reported from the geographical zone at 33°S. This zone is inhabited by species that are otherwise largely distributed in both northern and southern Chile, and thus can be characterized as a transition zone between those northern and southern Chilean inland water faunas. The southern extreme of Chile shows endemic species such as *Daphnia dadayana*, which inhabits shallow pools, whether permanent or ephemeral (Soto & De los Ríos, 2006; De los Ríos, 2008; De los Ríos et al., 2008). In contrast, for the large, deep lakes the species reported are similar to those found in northern (39-41°S) and central (41-48°S) Patagonian water bodies (De los Ríos & Soto, 2006; De los Ríos, 2008).

The species reported for Chile from the west side of the Andes mountains are low in number (Araya & Zúñiga, 1985; De los Ríos & Soto, 2007b) in comparison with those occurring on the east side of the Andes, mainly in Bolivia and Argentina (Olivier, 1962). This pattern, however, is similar to that reported for other planktonic crustaceans such as calanoid copepods (see next chapter; cf. also Bayly, 1992; Menu-Marque et al., 2000). In regard of the phenomenon observed, the mountains of the Andes chain would thus play an important role as a physical barrier, preventing the dispersion of species on both sides. Yet, the dispersal mechanism involving passive transport on migratory aquatic birds, would nonetheless allow colonization over long distances, eventually also crossing the Andes mountain ranges (De los Ríos & Zúñiga, 2000). In all, the marked differences found in the climate along the extended Chilean territory would definitely have to be acknowledged as a major cause for the presence of distinct biogeographical zones, with concurrent differences in biotic composition, all along the geographical range of the country (Luebert & Pliscoff, 2006; Vila et al., 2006).

A CHECKLIST OF THE COPEPODA: CALANOIDA

Chapter summary. — Calanoid copepods constitute the main component of the zooplankton assemblages found in Chilean inland waters. They are represented by three genera: *Boeckella*, *Parabroteas*, and *Tumeodiaptomus*. The genus *Boeckella* is chiefly represented by three species, i.e., first of all *B. gracilipes*, distributed in practically all inland fresh waters of the Chilean continental territory, primarily between 38 and 44°S. Another important species is *B. poopoensis* that inhabits saline lakes (5-90 g/L) of northern Chile (14-27°S), and the third species is *B. michaelseni*, inhabiting inland fresh waters from 44 to 54°S. The genus *Tumeodiaptomus* is represented by *T. diabolicus* that is dominant between 32 and 48°S. Finally, the genus *Parabroteas* has one species, *P. sarsi*, mainly abundant in shallow ponds in the range of 44-54°S. Ecological and biogeographical topics on these faunal elements are discussed below.

Introduction

Calanoid copepods are abundant in Chilean continental waters, due mainly to the frequently occurring conditions of oligotrophy, as reported for central (33-41°S) and southern (41-51°S) inland waters (Soto & Zúñiga, 1991; De los Ríos & Soto, 2006). However, calanoid copepods are also abundant in zooplankton assemblages in the northern part of the country, were salinity is the main regulating factor in the structure of zooplankton communities (De los Ríos & Crespo, 2004; De los Ríos, 2005). Finally, in southern Chile there is a different situation, because there exist numerous shallow ponds with markedly different conductivities and consequently with distinctly different trophic status (Soto et al., 1994; De los Ríos et al., 2008a, b). Here, the calanoid copepods are abundant in conditions of oligotrophy or high conductivity (Soto & De los Ríos, 2006). According to the literature, perhaps the most representative species is *B. gracilipes*, which has been reported for a wide range of shallow water bodies, distributed over a vast geographical area between 18 and 51°S (Villalobos & Zúñiga, 1991; Bayly, 1992a, b). Yet, it can be characterized as abundant only between 33 and 44°S

(Bayly, 1992; Menu-Marque et al., 2000). Another representative species is *Tumeodiaptomus diabolicus*, reported from 32 to 42°S (Soto & Zúñiga, 1991; Villalobos et al., 2003). Also worth mentioning is *B. poopoensis*, mainly inhabiting saline lakes of northern Chile (De los Ríos & Crespo, 2004) and of the surrounding countries (Menu-Marque et al., 2000; De los Ríos & Contreras, 2005). Finally, in the extreme south, the representative species are *B. michaelseni* and *Parabroteas sarsi*, both of which are frequently found from 44°S to the south. The species first mentioned is dominant in large lakes and shallow ponds, whereas the second mainly dominates in shallow ponds (De los Ríos, 2008). The present chapter provides a checklist of the calanoid copepods currently known from Chilean inland waters.

List of Calanoida reported from Chilean inland waters

Class MAXILLOPODA E. Dahl, 1956
- Subclass COPEPODA H. Milne-Edwards, 1840
  - Order CALANOIDA G. O. Sars, 1903
    - Family CENTROPAGIDAE Giesbrecht, 1893
      - Genus *Boeckella* De Guerne & Richard, 1889
        - *Boeckella bergi* Richard, 1897
        - *Boeckella brasiliensis* (Lubbock, 1855)
        - *Boeckella brevicaudata* (Brady, 1875)
        - *Boeckella calcaris* (Harding, 1955)
        - *Boeckella gibbosa* (Brehm, 1935)
        - *Boeckella gracilipes* Daday, 1901
        - *Boeckella gracilis* (Daday, 1902)
        - *Boeckella meteoris* Kiefer, 1928
        - *Boeckella michaelseni* (Mrázek, 1901)
        - *Boeckella occidentalis* (Marsh, 1906)
        - *Boeckella poopoensis* Marsh, 1906
        - *Boeckella poppei* (Mrázek, 1901)
      - Genus *Parabroteas* Mrázek, 1901
        - *Parabroteas sarsi* (Ekman, 1905)
    - Family DIAPTOMIDAE Baird, 1905
      - Genus *Tumeodiaptomus* Dussart, 1979
        - *Tumeodiaptomus diabolicus* (Brehm, 1935) [= *Diaptomus diabolicus* Brehm, 1935]

Records of species reported

Family CENTROPAGIDAE

Boeckella bergi Richard, 1897 (fig. 12): Aculeo lagoon (33°50′S 70°56′W) (Brehm, 1936).

Boeckella brasiliensis (Lubbock, 1855) (fig. 12): Balmaceda pools (45°53′S 71°40′W), Monserrat lagoon (51°07′S 72°47′W) (De los Ríos, 2005); Redonda lagoon (51°01′S

72°52′W), Larga lagoon (51°01′S 72°52′W), Jovito lagoon (51°02′S 72°54′W) (Bayly, 1992b).

Boeckella brevicaudata (Brady, 1875) (fig. 12): Balmaceda (45°53′S 71°40′W), Kon Aikén (52°50′S 71°10′W) (De los Ríos, 2005); De los Patos Bravos lagoon (53°09′S 70°57′W) (Mrázek, 1901).

Boeckella calcaris (Harding, 1955) [fig. 15]: Salar de Surire (18°15′S 69°07′W) (De los Ríos & Contreras, 2005).

Boeckella gibbosa (Brehm, 1935) (fig. 12): Pond close to Todos los Santos lake (41°08′S 71°56′W) (Brehm, 1935c); Negra lagoon (33°36′S 70°07′W) (Araya & Zúñiga, 1985); De los Indios lagoon (33°40′S 70°08′W) (Brehm, 1936a), Lo Encañado lagoon (33°40′S 70°08′W) (Brehm, 1935d).

Boeckella gracilipes Daday, 1901 (fig. 13): Parinacota lagoon (17°12′S 69°34′W) (Villalobos & Zúñiga, 1991), Cotacotani lagoon (18°14′S 69°13′W), Chungara lake (18°15′S 69°10′W), Del Inca lagoon (32°48′S 70°08′W), Negra lagoon (33°36′S 70°07′W), Panguipulli lake (39°43′S 72°15′W), Neltume lake (39°47′S 71°58′W), Pirihueico lake (39°50′S 71°49′W) (Araya & Zúñiga, 1985); Galletue lake (38°41′S 71°16′W), Icalma lake (38°49′S 71°17′W) (Parra et al., 1991); Caburgua lake (39°07′S 71°46′W) (Zúñiga, 1988); Pichilafquen lake (39°13′S 72°40′W) (Thomasson, 1963); Riñihue lake (39°50′S 72°19′W) (Zúñiga & Domínguez, 1977); Villarrica lake (39°18′S 72°06′W), Calafquen lake (39°31′S 72°13′W), Pellaifa lake (39°36′S 71°58′W), Ranco lake (40°13′S 72°25′W), Puyehue lake (40°40′S 72°28′W), Rupanco lake (40°50′S 72°31′W), Llanquihue lake (41°08′S 72°49′W), Todos los Santos lake (41°08′S 71°56′W) (Löffler, 1962); Chapo lake (41°27′S 72°31′W) (Soto & Zúñiga, 1991); Elizade lake (45°44′S 72°20′W), La Paloma lake (45°46′S 72°11′W) (Araya & Zúñiga, 1985); Cisnes lagoon (51°01′S 72°52′W), Redonda lagoon (51°01′S 72°52′W), Larga lagoon (51°01′S 72°52′W) (Bayly, 1992); Balmaceda pools (45°53′S 71°40′W) (De los Ríos, 2005); Don Alvaro lagoon (51°01′S 72°52′W), Guanaco lagoon (51°01′S 72°50′W), Paso lagoon (51°02′S 72°55′W) (De los Ríos, 2005); Sarmiento lake (51°03′S 72°47′W) (Campos et al., 1994a); Del Toro lake (51°12′S 72°45′W) (Campos et al., 1994b).

Boeckella gracilis (Daday, 1902) (fig. 13): Chungara lake (18°15′S 69°10′W) (Andrew et al., 1989); Riñihue lake (39°50′S 72°19′W) (Zúñiga & Domínguez, 1978); Calbuco lagoon (41°16′S 72°32′W) (Löffler, 1962); Mausa (41°27′S 72°58′W) (Brehm, 1937); lagoons at Cañi Park (39°15′S 79°43′W) (De los Ríos & Roa, 2010); Tinquilco (39°10′S 71°43′W), Verde lake (38°40′S 71°37′W), Marimenuco pools (38°40′S 71°05′W), Puaucho pools (38°57′S 73°09′-73°20′W) (P. De los Ríos et al., unpubl. data).

Boeckella meteoris Kiefer, 1928 (fig. 13): Cisnes lagoon (51°01′S 72°52′W) (Bayly, 1992), Cajon de Plomo (33°07′S 70°08′W) (Brehm, 1935c).

Boeckella michaelseni (Mrázek, 1901) (fig. 13): Polux lake (45°43′S 71°53′W), Elizade lake (45°44′S 72°20′W), General Carrera lake (45°50′S 72°00′W), Chiguay lake (45°56′S 71°50′W), Lynch lake (48°33′S 75°34′W) (Araya & Zúñiga, 1985); Jovito lagoon (51°02′S 72°54′W), Redonda lagoon (51°01′S 72°52′W) (Bayly, 1992b); Isidoro lagoon (50°57′S 72°53′W), Monserrat lagoon (51°07′S 72°47′W), Vega del Toro pools (51°07′S 71°40′W) (De los Ríos, 2005); Juncos lagoon (51°01′S 72°52′W), Pehoe lake (51°03′S 73°04′W), Norsdenkjold lake (51°03′S 72°58′W) (Soto & De los Ríos, 2006); Sarmiento lake (51°03′S 72°47′W) (Campos et al., 1994a); Del Toro lake (51°12′S 72°45′W) (Campos et al., 1994b).

Boeckella occidentalis Marsh, 1906 (fig. 14): Cotacotani lagoon (18°14′S 69°13′W), Chungara lake (18°15′S 69°10′W) (Bayly, 1992).

Boeckella poopoensis Marsh, 1906 (fig. 14): Calientes I lagoon (23°08′S 67°25′W), Calientes II lagoon (23°31′S 67°34′W), Calientes III lagoon (25°00′S 68°38′W) (Bayly, 1992); Chiuchiu lagoon (22°20′S 68°40′W) (Brehm, 1935a); Gemela Este lagoon (23°14′S 68°14′W),

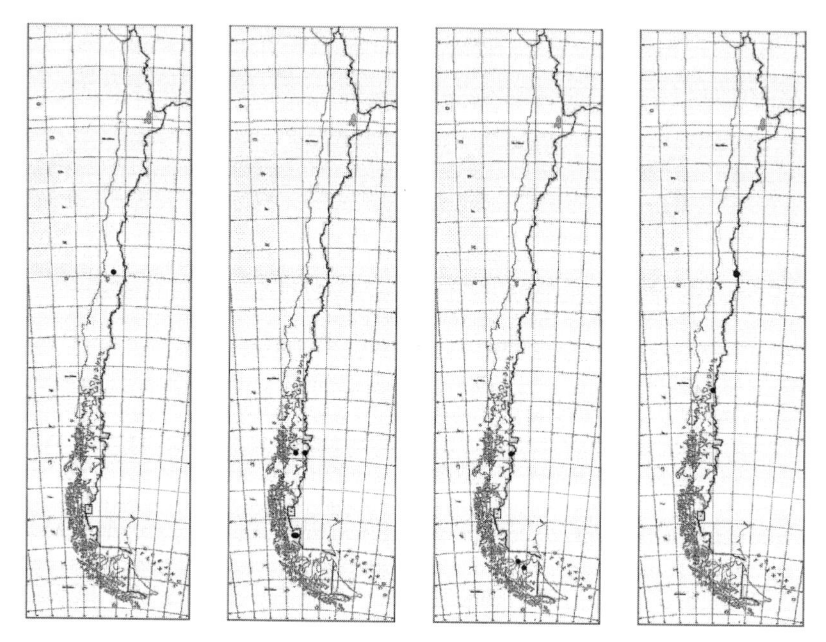

Fig. 12. Records of the following species (left to right): *Boeckella bergi* Richard, 1897; *Boeckella brasiliensis* (Lubbock, 1855); *Boeckella brevicaudata* (Brady, 1875); *Boeckella gibbosa* (Brehm, 1935).

Gemela Oeste lagoon (23°16′S 68°14′W), Miscanti lagoon (23°43′S 67°48′W), Miniques lagoon (23°43′S 67°48′W), Capur lagoon (23°54′S 67°48′W), Santa Rosa lagoon (27°05′S 69°10′W) (De los Ríos & Crespo, 2004).

Boeckella poppei (Mrázek, 1901) (fig. 14): Balmaceda pools (45°53′S 71°40′W), Isidoro lagoon (50°57′S 72°53′W), Don Alvaro lagoon (51°01′S 72°52′W), Guanaco lagoon (51°01′S 72°50′W), Monserrat lagoon (51°07′S 72°47′W), Vega del Toro pools (51°07′S 71°40'W), Kon Aikén pools (52°50′S 71°10′W), Porvenir pool (53°17′S 70°19′W) (De los Ríos, 2005); Cisnes lagoon (51°01′S 72°52′W), Redonda lagoon (51°01′S 72°52′W), Larga lagoon (51°01′S 72°52′W), Paso lagoon (51°02′S 72°55′W), Jovito lagoon (51°02′S 72°54′W) (Soto & De los Ríos, 2006).

Parabroteas sarsi (Ekman, 1905) (fig. 14): Los Palos lagoon (45°19′S 72°42′W), Riesco lake (45°46′S 72°20′W) (Villalobos, 1999); Chiguay lagoon (45°56′S 71°50′W), Elizalde lake (45°45′S 72°25′W) (Araya & Zúñiga, 1985); Balmaceda pools (45°53′S 71°40′W), Guanaco lagoon (51°01′S 72°50′W), Don Alvaro lagoon (51°01′S 72°52′W), Vega del Toro pools (51°07′S 71°40′W), Kon Aikén pools (52°50′S 71°10′W), Kon Aikén pools (52°50′S 70°50′W), Porvenir pool (53°17′S 70°19′W) (De los Ríos, 2005); Redonda lagoon (51°01′S 72°52′W), Larga lagoon (51°01′S 72°52′W), Cisnes lagoon (51°01′S 72°52′W) (Soto & De los Ríos, 2006); Monte lagoon and De los Patos Bravos lagoons (53°09′S 70°57′W) (Mrázek, 1901).

Family DIAPTOMIDAE

Tumeodiaptomus diabolicus (Brehm, 1935) (fig. 15): Runge reservoir (33°00′S 71°29′W), Peñuelas reservoir (33°10′S 71°29′W) (Schmid-Araya & Zúñiga, 1992); Valdivia (39°49′S

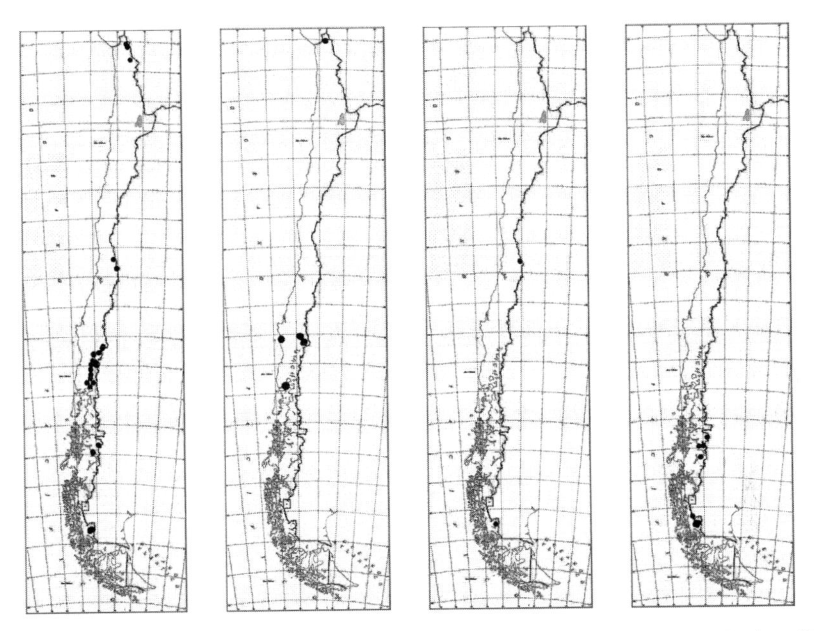

Fig. 13. Records of the following species (left to right): *Boeckella gracilipes* Daday, 1901; *Boeckella gracilis* (Daday, 1902); *Boeckella meteoris* Kiefer, 1928; *Boeckella michaelseni* (Mrázek, 1901).

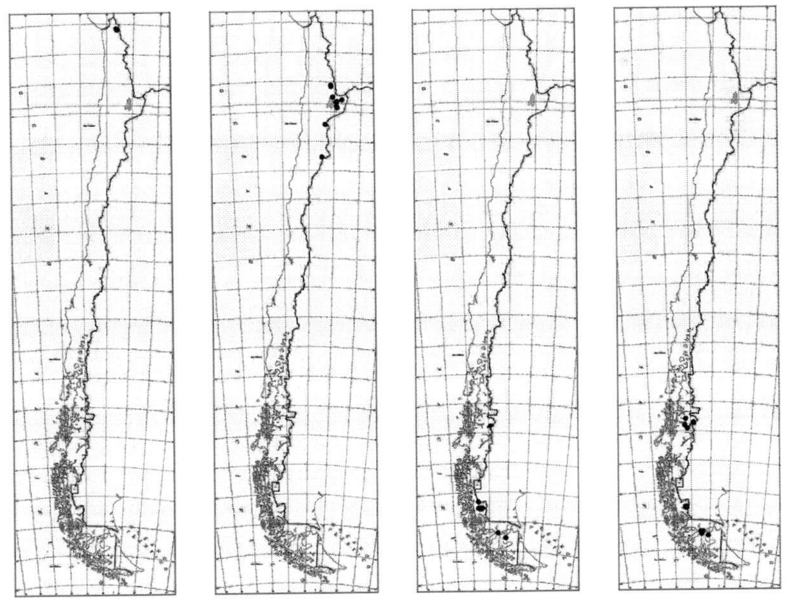

Fig. 14. Records of the following species (left to right): *Boeckella occidentalis* (Marsh, 1906); *Boeckella poopoensis* Marsh, 1906; *Boeckella poppei* (Mrázek, 1901); *Parabroteas sarsi* (Ekman, 1905).

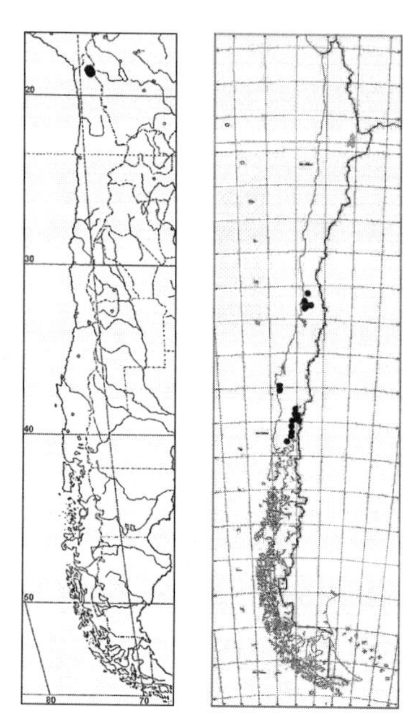

Fig. 15. Records of the following species (left to right: *Boeckella calcaris* (Harding, 1955); *Tumeodiaptomus diabolicus* (Brehm, 1935)).

73°15′W) (Brehm, 1935b); Pichilafquen lake (39°13′S 72°40′W), Calafquen lake (39°31′S 72°13′W), Pellaifa lake (39°36′S 71°58′W), Riñihue lake (39°50′S 72°19′W), Ranco lake (40°13′S 72°25′W), Puyehue lake (40°40′S 72°28′W) (Löffler, 1962); Catapilco reservoir (32°38′S 71°27′W), El Peral (33°30′S 71°35′W), Lanalhue lake (37°55′S 73°19′W), Lleulleu lake (38°08′S 73°19′W), Neltume lake (39°47′S 71°59′W), Pirihueico lake (39°56′S 71°48′W), Rupanco lake (40°50′S 72°31′W) (Araya & Zúñiga, 1985); Rapel reservoir (34°10′S 71°29′W) (Zúñiga & Araya, 1982); Sauzalito lagoon (33°00′S 71°32′W), Peñuelas lagoon (33°09′S 71°32′W), Orozco reservoir (33°14′S 71°25′W), Casablanca (33°18′S 71°24′W), Plateado reservoir (33°04′S 71°39′W), Villarrica lake (39°18′S 72°06′W), Llanquihue lake (41°08′S 72°49′W) (Zúñiga, 1975).

Discussion

The results reveal that some species are restricted to northern Chile, mainly occurring in saline and subsaline water bodies located in the Andes (Bayly, 1993; De los Ríos & Crespo, 2004; De los Ríos, 2005; De los Ríos & Contreras, 2005), such as *Boeckella occidentalis* and *B. poopoensis*. In this region it is also possible to find the widespread species *B. gracilipes* that is distributed over the entire Chilean continental territory. The species reported for northern Chile agree with literature descriptions for the surrounding zones in the Andes mountains of Argentina, Bolivia, and Peru (Menu-Marque et al., 2000). Nevertheless, from Chile only few species have been reported in comparison with the directly comparable (and more or less adjoining) areas in Argentina and Bolivia (Menu-Marque et al., 2000). A probable factor contributing to this phenomenon would obviously be constituted by the barrier formed by the Andes mountains, that would hamper the exchange of species between Chile and its surrounding regions (Gajardo et al., 1998; De los Ríos & Zúñiga, 2000; De los Ríos & Contreras, 2005).

In contrast to this observation, there are species that have nonetheless succeeded in extending their distributional range to encompass both sides of the Andes range, of which perhaps the most representative is the halophilic copepod, *Boeckella poopoensis*. This species is widespread in saline waterbodies of the whole of South America (Menu-Marque et al., 2000; Echaniz et al., 2006a) and can tolerate salinity levels ranging from 5 up to 90 g/L (Hurlbert et al., 1984, 1986; Bayly, 1993; Wiliams et al., 1995; De los Ríos & Crespo, 2004). It can even be the exclusive element in crustacean zooplankton assemblages at salinities between 20 and 90 g/L (Hurlbert et al., 1984, 1986; Wiliams et al., 1995; Zúñiga et al., 1999; De los Ríos & Crespo, 2004).

A different situation is to be observed in central Chile (33-37°S), where the representative species is the diaptomid, *Tumeodiaptomus diabolicus*, which is distributed in water bodies such as lagoons and reservoirs in valleys at moderate altitudes (Zúñiga, 1975; Araya & Zúñiga, 1985; Schmid-Araya & Zúñiga, 1992). Nevertheless, it is possible to find also here *Boeckella gracilipes*, which is reported from water bodies located in the mountains of the Andes. These observations would be corroborated by the data presented by Zúñiga (1975), who described *T. diabolicus* as being located in water bodies at lower altitudes in central Chile, because this species does not tolerate low temperatures. Perhaps the presence of *Boeckella bergi* in Aculeo lagoon is an exception, i.e., the presence of *Boeckella* in valley waters at moderate altitudes, where *T. diabolicus* usually is dominant. A possible cause would be associated with the deliberate introduction of fishes, i.e., the Argentinean silverside (*Odontesthes bonariensis* (Valenciennes in Cuvier & Valenciennes, 1832); cf. S. Menu-Marque, pers. comm.), i.e., the copepods having been accidentally introduced with those fish. Unfortunately, there are no studies available on planktonic crustaceans in waters located between 27 and 33°S. Likewise, studies between 34 and 38°S are scant, and restricted to only three sites, whereas there occur many lagoons and small reservoirs in that area that have not been studied yet.

The copepods occurring south of 38°S have been well studied through the comprehensive limnological studies that have been made of the water bodies in those regions (Domínguez & Zúñiga, 1979; Campos, 1984; Soto & Zúñiga, 1991; De los Ríos & Soto, 2006, 2007). The most intensively studied waters in these regions are the large, deep lakes located between 38 and 51°S, that are characterized by their oligotrophic or at least oligomesotrophic status (Soto & Zúñiga, 1991). Included in these are the so-called Araucanian lakes, located in the range of 38-41°S (Campos, 1984). The species reported from there are mainly *Tumeodiaptomus diabolicus*, distributed between 38 and 40°S, that can locally coexist with *Boeckella gracilipes* (cf. Soto & Zúñiga, 1991; De los Ríos & Soto, 2007). As for the latter species, this is a dominant element from 40° to approximately 45°S. Further south than 45°S, this species is known to be replaced by *B. michaelseni* (cf. Menu-Marque et al., 2000; De los Ríos & Soto, 2007; De los Ríos, 2008).

Also, in the zone of 38-42°S, there are two different groups of water bodies, the first being composed of the coastal lagoons and wetlands, in which *T. diabolicus* is present. This species has been reported at 41-42°S (De los Ríos, 2003; Villalobos et al., 2003), but more research in coastal wetlands and lagoons at 38-41°S will be required to understand the species' distribution as well as the actual existence of allopatry between *T. diabolicus* and *B. gracilipes*. The second group of inland waters consists of the numerous pristine mountain lakes, pools, and ponds located primarily inside protected areas with native forests between 39 and 45°S (Steinhart et al., 1999, 2002; De los Ríos et al., 2007). Those water bodies have not yet been studied systematically, due to serious problems in approaching and visiting these sites at higher altitudes in quite inaccessible terrain (De los Ríos et al., 2007). The few zooplankton studies that have been executed there report the presence of *B. gracilis* in lakes at 39°S (De los Ríos et al., 2007; De los Ríos & Romero, 2009), and it is probable that this species is distributed in high mountain lakes of northern Patagonia (38-41°S). Also here, however, it is necessary to do more studies in these ecosystems in order to acquire a more complete picture of the actual composition of the zooplankton assemblages thriving there.

The final zone is characterized by shallow ponds located in central and southern Patagonia between 45 and 53°S, which waters are either permanent or ephemeral, and show a wide conductivity range and a relatively high number of species of Crustacea (cf. Henríquez, 2004; De los Ríos, 2005, 2008), with in addition a remarkable share of endemism (Menu-Marque et al., 2000). The species reported are, within their group, the larger and more robust forms (Menu-Marque et al., 2000) and are mostly markedly pigmented (Villafañe et al., 2000). These are species that can tolerate a relatively wide range of conductivity, such as *Boeckella poppei* and *Parabroteas sarsi* (cf. De los Ríos & Contreras, 2005; De los Ríos & Rivera, 2007), and it is also possible to find here the halophilic copepod, *B. poopoensis* (cf. Menu-Marque et al., 2000; Soto & De los Ríos, 2006; Adamowicz et al., 2007). In addition, the widespread species, *B. gracilipes* and *B. michaelseni* can be found here (Menu-Marque et al., 2000; De los Ríos, 2008). According to the biogeographical analysis of Menu-Marque et al. (2000), this zone would be the dispersion zone of South American centropagids, because here occur species that have been reported both from South America and from the sub-Antarctic islands. Much more work needs to be done, however, to fully understand the distribution of the species that have, until now, been reported from these areas with low frequency, and that thus are in need of confirmation of their distribution, such as *Boeckella brevicaudata* and *B. gibbosa* (cf. Bayly, 1992; Menu-Marque et al., 2000).

A CHECKLIST OF THE COPEPODA: CYCLOPOIDA

Chapter summary. — The Cyclopoida may well be the most poorly studied copepods of Chilean inland waters, because the reports on the occurrence of species from this group need confirmation as regards the proper identification of those forms as well as on their spatial distribution. Moreover, only the cyclopoids of the large, deep Patagonian lakes (38-51°S) have been seriously investigated until now. On the basis of general limnological studies, however, we can state that the cyclopoids found show a low relative abundance in zooplankton assemblages. Likewise, the scarce ecological studies available would indicate that cyclopoids would be opportunistic predators, since they can graze on phytoplankton, feed on Protozoa, or predate on small multicellular zooplankton like rotifers, nauplii, copepodid instars, and the juvenile stages of cladocerans. This situation would be different in comparison to observations from the Northern Hemisphere that are strongly indicative of a role for cyclopoids as zooplankton predators. From a biogeographical point of view, there are endemic species as well as widespread species, but more studies are needed to confirm the actual extension of the ranges of the species reported. The same can be said with regard to the necessity to confirm the taxonomic identity of cyclopoid populations distributed mainly in northern water bodies (18-27°S) and in the shallow lagoons of southern Chile (45-54°S), as modern identification techniques may well reveal the existence of cryptic species, thus denying the widespread nature of some allegedly "cosmopolitan" forms.

Introduction

The cyclopoid copepods perhaps comprise the most poorly studied planktonic crustaceans of Chilean lakes, because this group is relatively scarce in comparison to other crustacean plankters such as calanoid copepods and cladocerans (Campos et al., 1982, 1983, 1987a, b, 1988, 1990, 1992a, b, 1994a, b; Soto et al., 1994; Wölfl, 1996; Villalobos, 1999; Villalobos et al., 2003a). The first review about cyclopoids is the report of Araya & Zúñiga (1985), and many of the species described in that paper were confirmed by Reid (1985). Nevertheless, some mistakes in taxonomic nomenclature were found that have been corrected during the last years (Locascio de Mitrovich & Menu-Marque, 2001; Pilati & Menu-Marque, 2003; Gutiérrez-Aguirre et al., 2006).

List of Cyclopoida reported from Chilean inland waters

Class MAXILLOPODA E. Dahl, 1956
- Subclass COPEPODA H. Milne Edwards, 1840
  - Order CYCLOPOIDA Burmeister, 1834
    - Family CYCLOPIDAE Dana, 1846
      - Genus *Acanthocyclops* Kiefer, 1927
        - *Acanthocyclops michaelseni* (Mrázek, 1901)
        - *Acanthocyclops vernalis* (Fischer, 1853)
      - Genus *Diacyclops* Kiefer, 1927
        - *Diacyclops andinus* Locascio de Mitrovich & Menu-Marque, 2001
      - Genus *Eucyclops* Claus, 1893
        - *Eucyclops ensifer* Kiefer, 1936
        - *Eucyclops serrulatus* (Fischer, 1851)
      - Genus *Macrocyclops* Claus, 1893
        - *Macrocyclops albidus* (Jurine, 1820)
      - Genus *Mesocyclops* G. O. Sars, 1914
        - *Mesocyclops araucanus* Löffler, 1962 (= *Mesocyclops longisetus araucanus* Löffler, 1962)
        - *M. longisetus* (Thiébaud, 1912)
      - Genus *Metacyclops* Kiefer, 1927
        - *Metacyclops mendocinus* (Wierzejski, 1892)
      - Genus *Microcyclops* Claus, 1893
        - *Microcyclops anceps* (Richard, 1897)
      - Genus *Paracyclops* Claus, 1893
        - *Paracyclops fimbriatus chiltoni* (Thomson, 1882)
      - Genus *Tropocyclops* Kiefer, 1927
        - *Tropocyclops prasinus meridionalis* Kiefer, 1927 (= *T. meridionalis* Kiefer, 1927)

Records of species reported

Family CYCLOPIDAE

Acantocyclops michaelseni (Mrázek, 1901) (fig. 16): Fagnano lake (54°31′S 68°43′W) (Mrázek, 1901).

Acantocyclops vernalis (Fischer, 1853) (fig. 16): Pirihueico lake (39°56′S 71°48′W) (Araya & Zúñiga, 1985); Chapo lake (41°27′S 72°31′W) (Villalobos et al., 2003).

Diacyclops andinus Locascio de Mitrovich & Menu-Marque, 2001 [= *Diacyclops bisetosus* (Rehberg, 1880)] (fig. 16): Chungará lake (18°15′S 69°10′W) (Araya & Zúñiga, 1985).

Eucyclops ensifer Kiefer, 1936 (fig. 16): Larga lagoon (51°02′S 72°55′W) (Kiefer, 1936).

Eucyclops serrulatus (Fischer, 1851) (fig. 17): Villarrica lake (39°16′S 72°07′W), Quillelhue lake (39°33′S 71°32′W), Puyehue lake (40°39′S 72°30′W), Bonita lagoon (40°53′S 72°52′W) (Löffler, 1962); Inca lagoon (32°49′S 70°09′W), Riñihue (39°49′S 72°19′W), Polux lake (45°43′S 71°53′W), Chiguay lake (45°56′S 71°50′W) (Araya & Zúñiga, 1985); Llanquihue lake (41°07′S 72°50′W) (Zúñiga & Domínguez, 1978).

Macrocyclops albidus (Jurine, 1820) (fig. 17): Quilpué (33°07′S 71°14′W), Valdivia (39°49′S 73°15′W) (Mrázek, 1901); Puyehue lake (40°39′S 72°30′W), Llanquihue lake (41°07′S 72°50′W) (Löffler, 1962); Pellaifa lake (39°30′S 71°57′W) (Zúñiga & Domínguez, 1977).

Mesocyclops araucanus Löffler, 1962 [= *Mesocyclops longisetus araucanus* Löffler, 1962, cf. Campos et al., 1974] (fig. 17): Calafquén lake (39°31′S 72°08′W), Pellaifa lake (39°30′S 71°57′W), Riñihue lake (39°49′S 72°19′W), Ranco lake (40°12′S 72°22′W), Puyehue lake (40°39′S 72°30′W), Rupanco lake (40°49′S, 72°30′W), Llanquihue lake (41°07′S 72°50′W), Todos los Santos lake (41°46′S 73°15′W), Bonita lagoon (40°53′S 72°52′W) (Löffler, 1962); Caburgua lake (39°07′S 71°47′W), Panguipulli lake (39°41′S 72°15′W), Pirihueico lake (39°56′S 71°48′W), General Carrera lake (45°50′S 72°00′W) (Araya & Zúñiga, 1985); Sarmiento lake (51°03′S 72°47′W) (Campos et al., 1994a); Del Toro lake (51°12′S 72°45′W) (Campos et al., 1994b).

Mesocyclops longisetus (Thiébaud, 1912) (fig. 17): Rapel reservoir (34°10′S 71°29′W) (Zúñiga & Araya, 1982); Negra lagoon (33°39′S 70°08′W), Lanalhue lake (37°55′S 73°19′W), Lleulleu lake (38°08′S 73°19′W) (Araya & Zúñiga, 1982).

Metacyclops mendocinus (Wierzejski, 1892) (fig. 18): Catapilco reservoir (32°38′S 71°27′W), Runge reservoir (33°01′S 70°54′W), Orozco reservoir (33°14′S 71°25′W), Elizalde lake (45°44′S 72°20′W) (Araya & Zúñiga, 1985); Rapel reservoir (34°10′S 71°29′W) (Zúñiga & Araya, 1982); Pichilafquen lagoon (39°13′S 72°12′W) (Löffler, 1962).

Microcyclops anceps (Richard, 1897) (fig. 18): Peral (33°30′S 71°35′W), Negra lagoon (33°39′S 70°08′W) (Araya & Zúñiga, 1985).

Paracyclops fimbriatus chiltoni (Thomson, 1882) (fig. 18): Quilpué (33°07′S 71°14′W) (Mrázek, 1901); Juan Fernández Island (?) (Brehm, 1936); Villarrica lake (39°16′S 72°07′W), Puyehue lake (40°39′S 72°30′W), Llanquihue lake (41°07′S 72°50′W), Margarita island (41°06′S 72°17′W) (Löffler, 1962); Peñuelas lagoon (Araya & Zúñiga, 1985).

Tropocyclops prasinus meridionalis Kiefer, 1927 [= *T. meridionalis* Kiefer, 1927] (fig. 18): Villarrica lake (39°16′S 72°07′W), Margarita island (41°06′S 72°17′W), Pocuro (32°53′S 70°38′W), Quillelhue lagoon (39°33′S 71°32′W), Bonita lagoon (40°53′S 72°52′W), Del Inca lagoon (32°49′S 70°09′W) (Löffler, 1962); El Plateado reservoir (33°04′S 71°39′W), Pellaifa lake (39°30′S 71°57′W), Puyehue lake (40°39′S 72°30′W) (Zúñiga & Domínguez, 1977); Riñihue lake (39°49′S 72°19′W) (Zúñiga & Domínguez, 1978); Ranco lake (40°12′S 72°22′W) (Domínguez & Zúñiga, 1978); Runge reservoir (33°01′S 70°54′W), Peñuelas reservoir (33°09′S 71°32′W), Orozco reservoir (33°14′S 71°25′W), Yeso reservoir (33°39′S 70°07′W), Lanalhue lake (37°55′S 73°19′W), Lleulleu lake (38°08′S 73°19′W), Caburgua lake (39°07′S 71°47′W), Calafquén lake (39°31′S 72°08′W), Neltume lake (39°47′S 71°59′W), Panguipulli lake (39°41′S 72°15′W), Pirihueico lake (39°56′S 71°8′W), Atravezado lake (45°45′S 72°54′W), General Carrera lake (45°50′S 72°00′W), Chiguay lake (45°56′S 71°50′W), Riesco lake (45°39′S 72°20′W), Lynch lake (48°33′S 75°34′W) (Araya & Zúñiga, 1985); Foitzick lagoon (45°38′S 72°05′W) (De los Ríos, 2008); Los Palos lagoon (45°19′S 72°42′W), Escondida lagoon (45°49′S 72°40′W) (Villalobos, 1999); Chapo lake (41°27′S 72°31′W) (Villalobos et al., 2003), Elizalde lake (45°44′S 72°20′W) (De los Ríos & Soto, 2007); Sarmiento lake (51°03′S 72°47′W) (Campos et al., 1994a); Del Toro lake (51°12′S 72°45′W) (Campos et al., 1994b); Pehoe lake (51°03′S 73°04′W); Norsdenkjold lake (51°03′S 72°58′W) (Soto & De los Ríos, 2006).

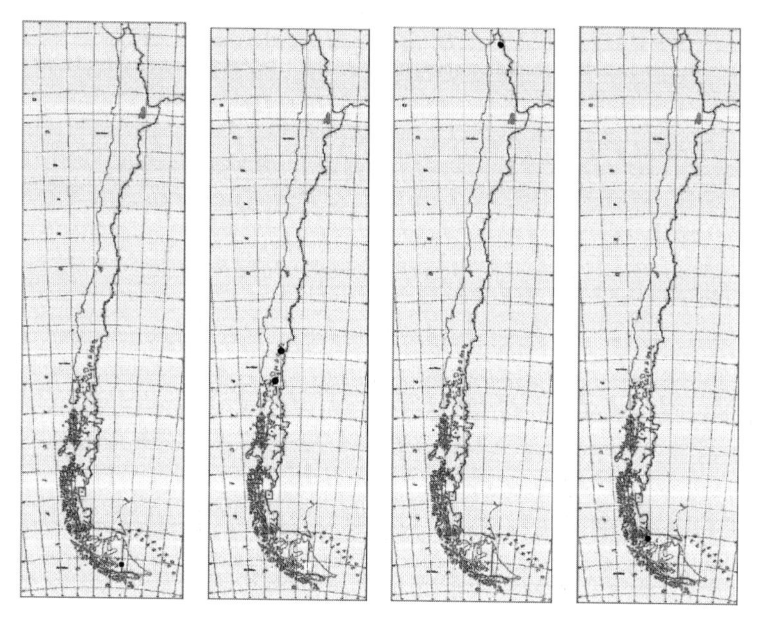

Fig. 16. Records of the following species (left to right): *Acanthocyclops michaelseni* (Mrázek, 1901); *Acanthocyclops vernalis* (Fischer, 1853); *Diacyclops andinus* Locascio de Mitrovich & Menu-Marque, 2001; *Eucyclops ensifer* Kiefer, 1936.

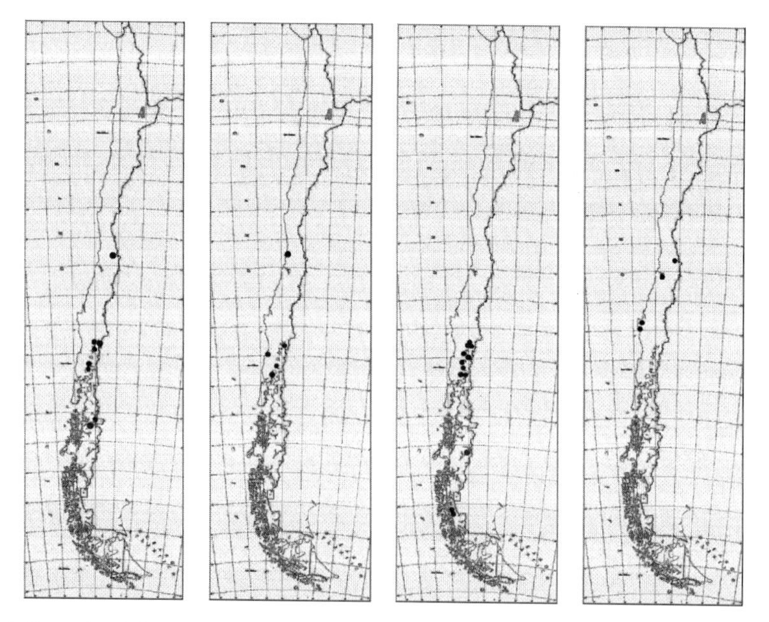

Fig. 17. Records of the following species (left to right): *Eucyclops serrulatus* (Fischer, 1851); *Macrocyclops albidus* (Jurine, 1820); *Mesocyclops araucanus* Löffler, 1962; *Mesocyclops longisetus* (Thiébaud, 1912).

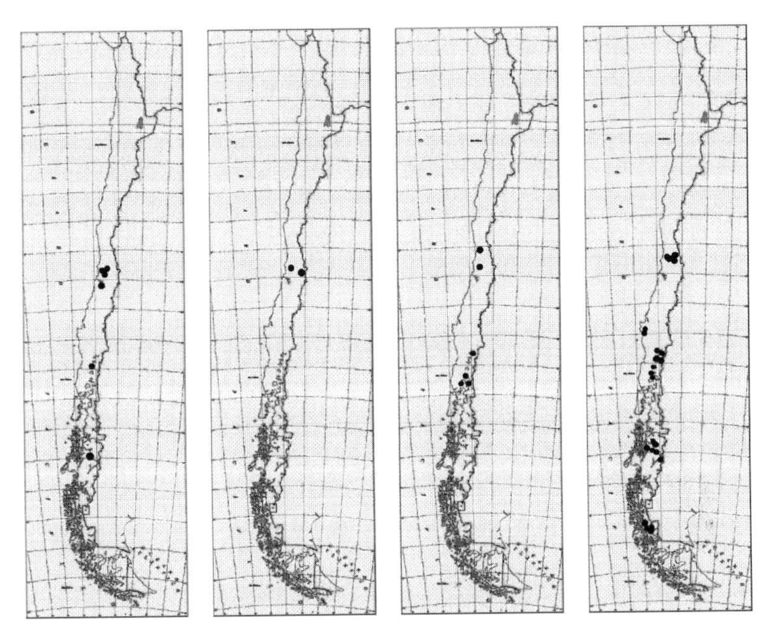

Fig. 18. Records of the following species (left to right): *Metacyclops mendocinus* (Wierzejski, 1892); *Microcyclops anceps* (Richard, 1897); *Paracyclops fimbriatus chiltoni* (Thomson, 1882); *Tropocyclops prasinus meridionalis* Kiefer, 1927.

Discussion

The information available about cyclopoids in Chilean inland waters shows the occurrence of a relatively low number of species in comparison with calanoids and cladocerans (De los Ríos & Soto, 2007; see also the two previous chapters herein). In addition, also the absolute abundance of these species in terms of numbers of individuals is low (Wölfl, 2007; Kamjunke et al., 2009), and similar observations have been described for Argentinean Patagonian lakes (Modenutti et al., 1998). From an ecological point of view, literature records indicate that cyclopoids are carnivorous, i.e., they predate on other groups in the zooplankton, such as rotifers, nauplii, and juvenile stages of cladocerans (Chang & Hanazato, 2003a, b, 2005a, b; Chang et al., 2004; Sakamoto & Hanazato, 2008). In regard of the ecosystems that have been described for the Chilean lakes where cyclopoids occur, their ecological status indicates that they can predate primarily on mixotrophic ciliates (Wölfl, 2007; Kamjunke et al., 2009). This was already presumed by Wölfl (1996), who mentioned that cyclopoids are opportunistic in their diet, because they are able to predate on protozoans or other, multicellular zooplankton organisms according to the specific conditions of their habitat.

In a biogeographical sense we find the presence of the widespread species *Mesocyclops longisetus*, originally described as being distributed all along the Chilean continental territory (Campos et al., 1982, 1983, 1987a, b, 1988, 1990, 1992a, b, 1994a, b; Araya & Zúñiga, 1985; Soto et al., 1994; Wölfl, 1996; Villalobos, 1999; Villalobos et al., 2003a). However, Pilati & Menu-Marque (2003) proposed that the populations of this species south of 38°S would belong to the species *M. araucanus*, originally described as a subspecies of *M. longisetus*, i.e., *M. longisetus araucanus* (cf. Löffler, 1962). The available literature does not allow to distinguish distinct biogeographical zones in Chile as regards the distribution of cyclopoid copepods, because the endemism observed in this group is low. The only allegedly endemic species are in need of confirmation of their (restricted) distribution: *Diacyclops andinus* that would inhabit northern Chile and northwestern Argentina (Locascio de Mitrovich & Menu-Marque, 2001), and *Acanthocyclops michaelseni*, which is supposed to inhabit the extreme southern parts of Argentina and Chile (Mrázek, 1901; Silva, 2008).

THE ECOLOGY OF THE CRUSTACEAN ZOOPLANKTON IN THE SALINE LAKES OF NORTHERN CHILE

Chapter summary. — The saline lakes of northern Chile are characterized by high salinity due to a combination of saline deposits and an arid climate. Salinity is the main regulating factor for crustacean zooplankton assemblages, because at salinities above 20 g/L only a single halophilic species is present: either *Boeckella poopoensis* at salinities between 20 and 90 g/L, or *Artemia franciscana* at salinities over 90 g/L. Although according to the literature both species can tolerate salinities between 20 and 90 g/L, field observations make clear that both species do not coexist. Although predation of *B. poopoensis* on *A. franciscana* nauplii might be postulated as a possible cause, there are no strong arguments for explaining the non-coexistence of these two species.

Introduction

The northern zone of Chile (18-27°S) is characterized by a markedly arid climate (Luebert & Pliscoff, 2006) and the presence of shallow water bodies associated with saline deposits of volcanic origin (Chong, 1888). These saline waters, however, have not been studied in detail due to access problems, and as a result the only information available derives from a few field studies (Zúñiga et al., 1991, 1994, 1999; De los Ríos & Crespo, 2004), as well as through recompilation of published information from short communications (De los Ríos, 2005; De los Ríos & Contreras, 2005). All the biotic elements in these communities are regulated by salinity: the primary producers are halophilic phytoplankton and/or bacteria that thus produce the feed for the grazer zooplankton (Zúñiga et al., 1991, 1994) or for aquatic birds (López, 1990). The zooplankton, in its turn, can be predated upon by zooplanktivorous birds (Hurlbert et al., 1984, 1986; López, 1990), or planktivorous fish that are, however, usually scarce in these water bodies (Vila et al., 2006), since they occur mainly at lower salinities (Keller & Soto, 1998).

The ecological descriptions at our disposal are chiefly based on field observations on the Bolivian and Peruvian Altiplano (Hurlbert et al., 1984, 1986; Williams et al., 1995). Thus, the first descriptions of northern Chilean saline lakes (18-27°S) were made with the aim of searching native *Artemia* populations, such within the framework of former research projects. The results confirmed that *Artemia* is the practically exclusive component in the zooplankton of those hypersaline lakes (Zúñiga et al., 1991, 1994, 1999; De los Ríos & Gajardo, 2004). Finally, recent studies described the ecology of other zooplankton elements, mainly halophilic copepods, as well as the species assemblages occurring in northern Chilean water bodies with a wider salinity range (De los Ríos & Crespo, 2004).

Geographical and general ecological features

The part of Chile between 18 and 27°S is denoted as the "Large Northern" zone (the regions Arica & Parinacota, Tarapacá, and Antofagasta) and its main characteristics are the arid climate with only scarce, shallow water bodies, and in some zones a practically total absence of precipitation (Aceituno, 1997; Luebert & Pliscoff, 2006). Similar characteristics are observed in a small zone between 26 and 27°S (the Atacama region), included in the so-called "Small Northern" zone, which forms a transition zone between the arid desert and a zone with a gradual increase in precipitation and hence the presence of small rivers, which are situated mainly, however, south of 27°S (Niemeyer & Cereceda, 1984; Luebert & Pliscoff, 2006). The main characteristic of the desert zone at 18-27°S is the presence of three permanent rivers (Lluta, San José, and Loa, fig. 19), and numerous small, ephemeral streams located in the Andes mountains. These originate from rain or from the emergence of underground waters, and they finish in small lagoons or saline deposits (Niemeyer & Cereceda, 1984); unfortunately, these rivers and small streams have either not, or only hardly been studied. Other important water bodies are the small, shallow lagoons that are mainly located in the Andes and these are associated with saline deposits called "salares" (Chong, 1988; Mühlhauser, 1997; Zúñiga et al., 1999; figs. 19, 20, 21). From the point of view of chemical limnology, a main characteristic of those salares is the presence of high sulphate concentrations in the brines, caused by the mineral composition of the deposits in the catchment basins (Chong, 1988; Richaser et al., 1999). These water bodies are the result of hydrogeological activity in the past, because in the Pleistocene the Andes mountains had a wet climate, with many wetlands, whereas during the Holocene this zone changed from wet to the present arid climate (Grosjean et al., 1995, 1996, 1997; Valero-Garcés et al., 1996; Kull & Grosjean, 1998). The catchment basins of these lagoons are characterized by

Fig. 19. Map of northern Chile with the locations of some representative water bodies.

the presence of xerophytic vegetation, i.e., mainly shrubs such as *Fabiana ramulosa* (Wedd.) Hunz. & Barboza, *F. denudata* Miers [Pflanzl.], *F. squamata* (Phill.), *F. bryoides* (Phill.), *Dipostephium meyenii* (Wedd.), *Pharastrepia lucida* (Meyen), *P. lepidophylla* (Wedd.), *P. quadrangularis* (Meyen), *Azorella compacta* (Phill.), *Festuca orthophylla* (Pilg.), *F. chrysophylla* (Phill.), *Pycnophyllum molle* (Remy), *Chuquiraga atacamensis* (Kuntze), *Mulinium crassifolium* (Phill.), *Urbania pappigera* (Phill.), *Artemisa copa* (Phill.), and *Stipa frigida* (Phill.) (cf. Luebert & Pliscoff, 2006).

The water bodies in northern Chile are nesting and feeding areas for aquatic birds such as the common gallinules (*Gallinula chloropus galeata* (Lichtenstein, 1818) and *Gallinula chloropus pauxilla* Bangs, 1915), crested ducks (*Lophonetta specularioides* (King, 1828)), the Chilean flamingo (*Phoenicopterus chilensis* Molina, 1782), the Andean flamingo (*Phoenicoparrus andinus* (Philippi, 1854)), James' flamingo (*Phoenicoparrus jamesi* (Sclater, 1886)), and Wilson's phalarope (*Phalaropus tricolor* (Vieillot, 1819)) (cf. Araya & Millie, 2005; Schlatter & Sielfield, 2006). The food of some water birds can be based on phytoplankton as is the case in the Andean flamingo (Hurlbert & Chang, 1963; Lopez, 1990, 1997), which probably can lead to competition between zooplankton and those aquatic birds. Other aquatic birds such as the Chilean flamingo and Wilson's phalarope predate on zooplankton (Hurlbert et al., 1984; López, 1990). These trophic interactions have been

Fig. 20. Photographs of: top, Miscanti lagoon; bottom, Salar de Llamara. [Photos P. R. De los Ríos-Escalante.]

Fig. 21. The Chilean flamingo, *Phoenicopterus chilensis* Molina, 1782, in: top, Salar de Atacama; and, bottom, Salar de Surire. [Photos P. R. De los Ríos-Escalante.]

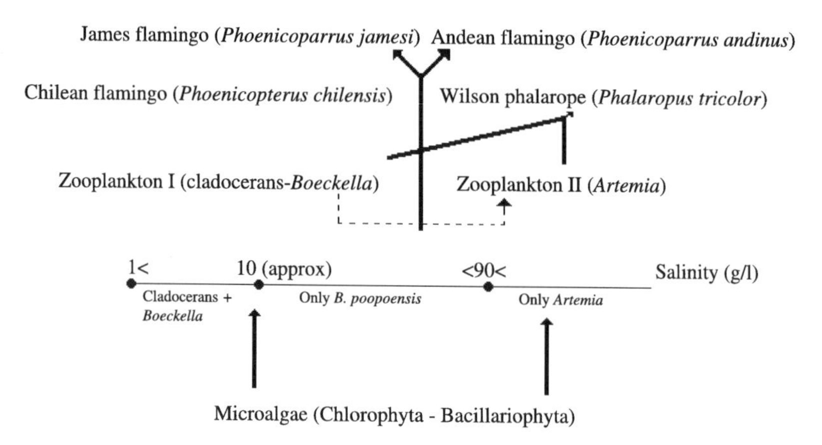

Fig. 22. Trophic interactions in fishless water bodies of northern Chile.

studied in basic field observations, made mainly at locations situated on the territories of Bolivia and Peru (fig. 22; Hurlbert, 1982; Hurlbert et al., 1984, 1986), and it is probable that these interactions are similar to those actually existing in the shallow lagoons in southern Patagonia (Soto, 1990). Then, in this scenario the zooplankton would play a key role as the main phytoplankton grazer, and in its turn as the main prey for the aquatic birds (fig. 22; Hurlbert et al., 1984, 1986; Bayly, 1995). The basis of the trophic webs are the diatoms and the hypersaline microalgae such as *Dunaliella salina* (Dunal) Teodoresco in the phytoplankton (Zúñiga et al., 1991; Lopez, 1997) as well as halophilic bacteria (Zúñiga et al., 1991, 1994; Campos, 1997; López, 1997; Demergasso et al., 2003), that would constitute the main prey for halophilic crustaceans, primarily *Artemia* (cf. Zúñiga et al., 1991, 1994; Gliwicz, 2003). Fish are scarcely present, and their occurrence is restricted to water bodies of low salinity (Keller & Soto, 1998). Those fish that do occur can be endemic species of genera such as *Orestias*, *Trichomycterus*, and *Odontesthes* (fig. 23; Dyer, 2001; Vila et al., 2006) as well as the introduced *Oncorhynchus mykiss* (Walbaum, 1792) (fig. 23; Wetzlar, 1979; Hurlbert et al., 1986; Vila et al., 2006). Unfortunately, no detailed studies have been published about the trophic interactions of fish and zooplankton, and only Hurlbert et al. (1986) described a potentially trophic interaction in shallow Andean lakes of Peru involving native and introduced fishes. Nevertheless, fish could well be opportunistic predators on zooplankton (Hurlbert et al., 1986).

From a biogeographical point of view this region has marked endemism (Menu-Marque et al., 2000; Vila et al., 2006), because of the presence of the high altitude ranges of the Andes mountains and the arid climate, which

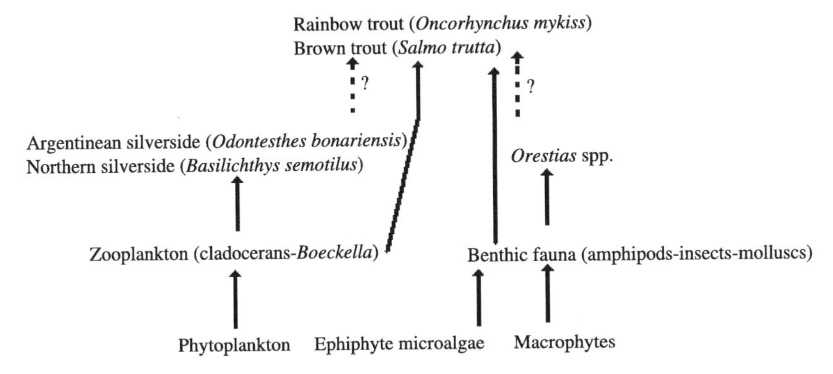

Fig. 23. Trophic interactions in water bodies with fishes in northern Chile.

together generate a specific environment that will allow only biota with specific adaptations to thrive there (Vila et al., 2006). The Andes chain has also put up a physical barrier against (easy) dispersion of the fauna on both sides of the mountains (Gajardo et al., 1998; De los Ríos & Zúñiga, 2000), which would explain the differences in crustacean zooplankton assemblages on both sides of the Andes range (Menu-Marque et al., 2000; De los Ríos & Contreras, 2005).

However, the water bodies in this region provide nesting, resting, and feeding areas for migratory water birds such as the Chilean flamingo, which has a long migratory route between the Altiplano and the island of Tierra del Fuego (Soto, 1990; Araya & Millie, 2001; Gibbons et al., 2007). If we consider that aquatic birds can be an important dispersal agent for crustacean cysts, for example *Artemia* (cf. Dana et al., 1988; Triantaphyllidis et al., 1998; Green et al., 2005), but also for other anostracans such as *Branchinecta* (cf. Brendonck, 1996), it would be probable that the Chilean flamingo and other migratory aquatic birds (Hurlbert & Keith, 1979) may explain the presence of widespread species such as *Boeckella poopoensis*. This species was reported along an extended geographical stretch between the Altiplano and southern Patagonia (Menu-Marque et al., 2000; Soto & De los Ríos, 2006), as well as from estuarine environments in southern Argentina (Hoffmeyer, 2004).

Crustacean assemblages

Due to the high salinity in the water bodies here at issue, the number of species in the zooplankton is low: five species at most are found (Hurlbert et al., 1984; Bayly, 1992a, b; Williams et al., 1995; De los Ríos, 2005), and these results are similar to those reported from other saline lakes (Williams,

1998). The descriptions of the assemblages in these lakes are also similar to the conditions in the surrounding lakes, where the halophilic copepod *Boeckella poopoensis* can be found at salinities below 90 g/L, whereas at salinities above 90 g/L only the anostracan *Artemia franciscana* is present (fig. 24; Hurlbert et al., 1986; Williams et al., 1995). For Chilean water bodies in general, a higher number of species has been observed at salinities between 5 g/L and 90 g/L (De los Ríos & Crespo, 2004), and then it is also possible to find two species of copepods coexisting (De los Ríos & Contreras, 2005). The wide salinity gradient encountered in the total of northern Chilean lakes allows to recognize a rather strong correlation between species richness and salinity, in comparison to saline or subsaline lakes in southern Patagonia (De los Ríos, 2005; Soto & De los Ríos, 2006).

In general, such a correlation is strongly inverse: the higher the salinity value, the lower the number of species present (table I; fig. 24). The main groups observed at low salinity are centropagid calanoid copepods, such

TABLE I

Salinity and percentages of *Boeckella poopoensis* Marsh and *Artemia franciscana* Kellogg, as representative species of northern Chilean saline lakes, with the percentages expressed as % of total halophilic Crustacea

Site	Salinity g/L	% of halophilic Crustacea	
		B. poopoensis	*A. franciscana*
Salar de Llamara	160.00	–	100
Gemela Este	41.00	54.1	–
Gemela Oeste	51.40	46.5	–
Cejas I	129.30	–	100
Cejas II	150.00	–	100
Cejas III	189.00	–	100
Tebenquiche	300.00	–	100
Chaxas	120.00	–	100
Miscanti	8.98	73.0*)	–
Miniques	9.79	50.0*)	–
Salar de Capur	3.40	66.0**)	–
Santa Rosa	8.00	99.0***)	–

Notes:

*) For Miscanti and Miniques the remaining percentage (27 and 50%, respectively) corresponds to *Daphnia* spp., *Alona pulchella* King, Chydoridae indet., and Cyclopoida indet. (cf. De los Ríos & Crespo, 2004).

**) For Capur the remaining percentage (34%) corresponds to Chydoridae indet. and Cyclopoida indet. (cf. De los Ríos & Crespo, 2004).

***) For Santa Rosa the remaining percentage (1%) corresponds to Chydoridae indet. (cf. De los Ríos & Crespo, 2004).

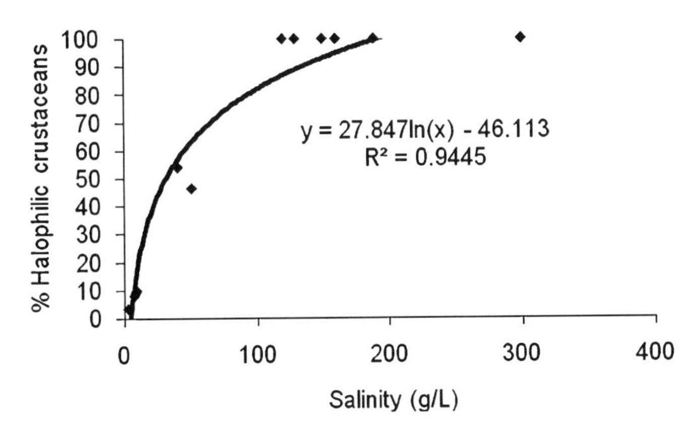

Fig. 24. Relationship between salinity and percentage of dominance of halophilic crustaceans (*Boeckella poopoensis* Marsh, 1906 or *Artemia franciscana* Kellogg, 1906) in water bodies of northern Chile.

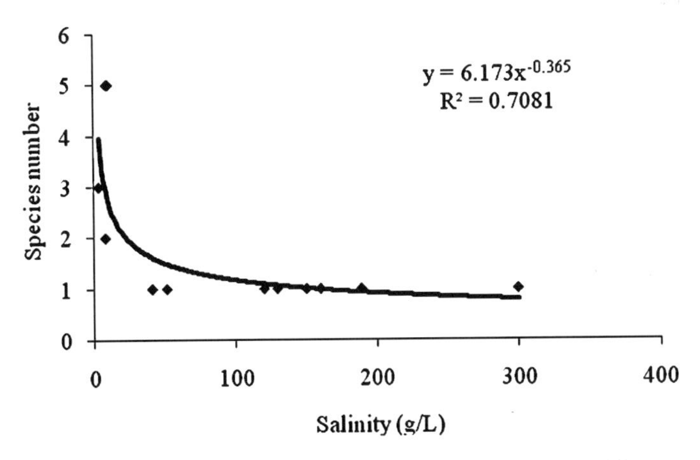

Fig. 25. Relationship between salinity and number of species in northern Chilean water bodies.

as *Boeckella gracilipes* and *B. occidentalis*, along with daphniids (*Daphnia pulex*), chydorid branchiopods (*Alona* sp.), and cyclopoid copepods (De los Ríos, 2005; De los Ríos & Contreras, 2005). As the salinity increases, the centropagids mentioned are replaced by their halophilic congener, *B. poopoensis*, which inhabits waters of salinities between 5 and 90 g/L (De los Ríos & Contreras, 2005). Also, in northern Chilean lakes *Boeckella calcaris* has been reported to coexist with *B. gracilipes* at low salinity, albeit only rarely (De los Ríos & Contreras, 2005), while it is also possible to find the species *B. meteoris* at low salinity values (Menu-Marque et al., 2000).

At salinities above 90 g/L, *Artemia franciscana* is the exclusive component of the zooplankton assemblages, and this species allegedly has a wide salinity tolerance of 15-200 g/L (Sorgeloos et al., 1986). Though with these tolerance ranges the coexistence of *A. franciscana* with *B. poopoensis* would theoretically be possible, both species do not coexist probably because *B. poopoensis* predates on the nauplii of *Artemia* (cf. Hurlbert et al., 1986; Hammer & Hurlbert, 1992; Williams, 1998). These results agree with observations on *B. poopoensis* for central Argentina, specifically in the province of La Pampa, where the species was described as having an omnivorous diet that includes phytoplankton, and in which also remnants of nauplii were found (Echaniz et al., 2006), while no coexistence with *Artemia* populations was observed there, either (Echaniz et al., 2006; Vignatti et al., 2007).

The inverse relationship between salinity and species number as noted above, thus culminates in eventually finding only one dominant species at the maximal increase of salinity (De los Ríos, 2005). Thus, centropagid copepods dominate at salinities up to 90 g/L, the most representative of these being the halophilic *Boeckella poopoensis*, occurring at values of 5-90 g/L (table II, figs. 24-25; De los Ríos & Crespo, 2004). At lower salinities the dominance of calanoids decreases and then cyclopoids and cladocerans can be found, although a dominance of cladocerans over calanoids has never been reported (table II, fig. 25; De los Ríos & Crespo, 2004).

These descriptions of species richness and dominance for northern Chilean lakes are in concordance with the conditions recorded for their Bolivian and Peruvian counterparts (Hurlbert et al., 1984, 1986; Williams et al., 1995). A similar pattern is also found for water bodies in Australia and New Zealand, where the local halophilic species of calanoid Copepoda are *Calamoecia clitellata* Bayly, 1962, and *C. salina* (Nicholls, 1944) that show salinity tolerances of 1-22 g/L, 6-132 g/L, and 7-195 g/L, respectively. Thus, the two species last mentioned have a remarkably higher salinity tolerance than the Chilean *B. poopoensis* (cf. Bayly, 1993). Another different condition is, that the Australian saline lakes show a wide variety of species, with at least three halophilic copepod genera that are widespread and with an apparently high dispersal capacity, due to both natural causes and anthropogenic intervention (Maly, 1996; Maly et al., 1997).

Unfortunately, there are no detailed accounts of community or population structure of these zooplankters in the saline lakes of northern Chile, although the studies of Zúñiga et al. (1991, 1994) do give at least some data. These authors described the population of *Artemia franciscana* in Salar de Atacama, and noted that at an increase of salinity a high adult mortality occurs, and when the salinity decreases again, the remaining adults would generate the new

TABLE II
Species reported from some northern Chilean saline lakes

	Salinity g/L	Species
Salar de Llamara	160.00	*Artemia franciscana* Kellogg
Gemela Este	41.00	*Boeckella poopoensis* Marsh
Gemela Oeste	51.40	*B. poopoensis*
Cejas I	129.30	*A. franciscana*
Cejas II	150.00	*A. franciscana*
Cejas III	189.00	*A. franciscana*
Tebenquiche	300.00	*A. franciscana*
Chaxas	120.00	*A. franciscana*
Miscanti	8.98	*B. poopoensis*
		Daphnia spp.
		Alona pulchella King
		Chydoridae indet.
		Cyclopoida indet.
Miniques	9.79	*B. poopoensis*
		Daphnia spp.
		A. pulchella
		Chydoridae indet.
		Cyclopoida indet
Salar de Capur	3.40	*B. poopoensis*
		Chydoridae indet.
		Cyclopoida indet.
Santa Rosa	8.00	*B. poopoensis*
		Chydoridae indet.

population [if not also new individuals hatch from dormant cysts]. A similar pattern of population behaviour was described for *Artemia monica* Verrill, 1869 (= *A. franciscana monica*), that shows a decrease in total body length, fecundity, and population growth when salinity increases (Dana & Lenz, 1986; Dana et al., 1993). For that same species, there are observations that the production of dormant cysts is directly associated with the reproductive activity of the previous generation (Dana et al., 1990).

These kinds of ecological behaviour were also described for the halophilic copepod, *Boeckella hamata* Brehm, 1928 that showed considerable mortality upon a strong increase of salinity (Hall & Burns, 2001a). This same species is able to produce resting eggs in an unfavourable environment, that will hatch as soon as conditions have become favourable again (Hall & Burns, 2001b). Similar observations have also been reported for other saline inland waters, such as for coastal lakes in Croatia (Kršinić et al., 2000). From northern Africa, the effects of variations in salinity on reproduction, survival, and growth of the halophilic copepod *Arctodiaptomus salinus* (Daday, 1885) (cf. Rokneddine, 2004b, 2005) and of the cladoceran *Moina salina* Daday, 1888 (cf. Rokneddine, 2004a) have been described. In comparable ecosystems in Australia and New Zealand, studies have been made on the decrease in survival and growth of *Daphnia carinata* King, 1853, when salinity increases (Hall & Burns, 2002). In the scenario there found, a salinity-based niche segregation of planktonic crustaceans appears to be established over a wide salinity gradient, with differences in salinity tolerance of the various freshwater species when the salinity increases (Pinder et al., 2005). In African saline lakes, LaBarbera & Kilham (1974) proposed a salinity tolerance for each species reported from the sites studied, based on the data found in the field.

If a similar scenario would be effective in the Chilean conditions, a segregation of crustacean zooplankters in subsaline, mesohaline, and saline water bodies would be indicated, as has indeed been described by De los Ríos & Contreras (2005). Another important factor that could affect copepod populations in saline and subsaline water bodies, is temperature. Extreme temperatures may cause mortality in late nauplius and early copepodid stages in the halophilic copepod *Arctodiaptomus salinus* (cf. Jimenez-Melero et al., 2007). Similar responses in embryonic and naupliar development at variable temperatures were observed for other diaptomid species, i.e., *Eudiaptomus gracilis* (G. O. Sars, 1862) and *E. graciloides* (Lilljeborg, 1888) (cf. Jimenez-Melero et al., 2005; Zeller et al., 2004).

The absence of halophilic crustaceans in hypersaline waters of northern Chile agrees with the reports of Bayly (1972), who mentioned that species of the genera *Boeckella* and *Calamoecia* are not present in coastal hypersaline waters, but occur in inland saline waters only. Also, the zooplankton assemblages observed in northern Chilean water bodies are different from those found in saline lakes of the province La Pampa in Argentina, where it is possible to find four daphniid species in a salinity range between 0.2 and 37.2 g/L (Echaniz & Vignatti, 1996; Paggi, 1996; Echaniz et al., 2006). The most halophilic of those is *Moina eugeniae* Olivier, 1954 that can tolerate salinities of 13.5-37.2 g/L (Echaniz & Vignatti, 1996). Another important difference is, that these Argentinean water bodies have a different complement of copepod species, that includes *Boeckella gracilipes*, *B. gracilis*, *Notodiaptomus incompositus* (Brian, 1925), and the widespread halophilic *Boeckella poopoensis* (cf. Pilati, 1997; Vignatti et al., 2007). It has also been observed that cyclopoids have a relative tolerance to sub-saline waters (0.2-5.2 g/L for *Acanthocyclops robustus* (G. O. Sars, 1863); 0.3-1.1 g/L for *Tropocyclops prasinus meridionalis*; 0.2-2.9 g/L for *Mesocyclops meridianus* (Kiefer, 1926); cf. Pilati, 1999) as well as for mesohaline water bodies (0.6-20.8 g/L for *Metacyclops mendocinus* and 0.4-12.8 g/L for *Microcyclops anceps*; cf. Pilati, 1999). This is a different picture in comparison to northern Chilean saline lakes, because these have no diaptomids in their sub-saline waters (De los Ríos & Crespo, 2004; De los Ríos & Contreras, 2005), and there is only one cyclopoid species present, i.e., *Diacyclops andinus* (cf. Locascio de Mitrovich & Menu-Marque, 2001).

In all, it thus seems inevitable to conclude that (a) ambient local conditions, and these (b) explicitly considered also in the context of their historical past, as well as, in addition, (c) the species that are, locally, actually available to populate a body of water, together determine the (type of) zooplankton community effectively present at any particular site.

THE ECOLOGY OF THE CRUSTACEAN ZOOPLANKTON IN CENTRAL CHILEAN SMALL LAKES AND RESERVOIRS

Chapter summary. — The central part of Chile has numerous reservoirs and small lakes located in agricultural valleys and plains. From the sites studied there with regard to crustacean plankton, mainly species occurrences have been reported, and the few ecological studies revealed the role of the trophic status of those water bodies in determining the zooplankton assemblage locally present. Many of these waters are eutrophic, due to the nutrient input from agricultural activities in the vicinity of their catchment basins. The zooplankton assemblages in this region are characterized by the presence of small cladocerans, mainly *Ceriodaphnia dubia*, *Moina micrura*, and *Neobosmina chilensis* [now referred to as *Eubosmina hagmanni*], and by the occurrence of the calanoid, *Tumeodiaptomus diabolicus*. An example of these kinds of lakes can well be the Nahuelbutan Lakes, located in a coastal mountain zone with a wide gradient of trophic status, where it is possible to find eutrophic lakes located close to urban zones, and mesotrophic lakes located in zones with agricultural and/or silvicultural activities.

Introduction

In the central region of Chile we find many agricultural valleys with rivers fed by a mixed regime, viz., the rains in winter and the melting Andean mountain snow in summer (Niemeyer & Cereceda, 1984; Montecino et al., in press). This zone has a Mediterranean climate, with marked seasons: the well-known sequence of rather dry, warm summers and moist, mild winters (Luebert & Plistoff, 2006). From an ecological point of view, this zone harbours a series of markedly fragmented habitats, because (large) parts of the native forest have been replaced by human agricultural activities (Grey & Bustamante-Sánchez, 2006), with the predictable consequences for those native species that can still be reported from these areas (Bustamante et al., 2006; Rau et al., 2006). The limnological studies performed are mainly reporting occurrences of species (Domínguez, 1973; Domínguez & Zúñiga, 1976; Zúñiga & Domínguez, 1977, 1978; Zúñiga & Araya, 1982; Araya &

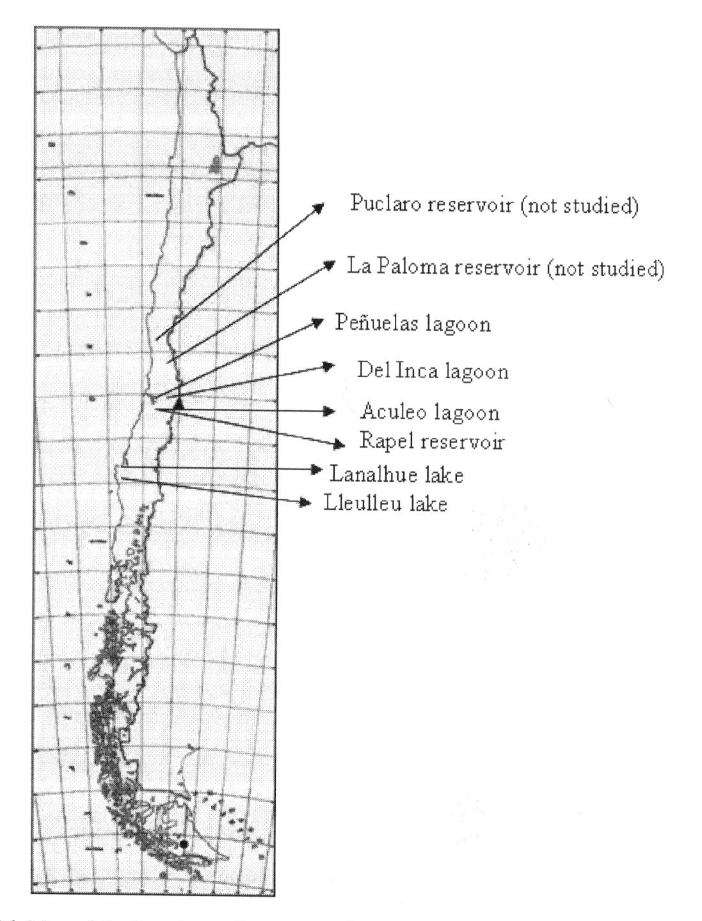

Fig. 26. Map of the locations of representative central Chilean lagoons and reservoirs.

Zúñiga, 1985). The few ecological studies done have established the existence of eutrophic lagoons and reservoirs (Cabrera & Montecino, 1987; Mühlhauser & Vila, 1987; Schmid-Araya & Zúñiga, 1992; Ramos et al., 1998; Muñoz et al., 2001; Ramos-Jiliberto & Zúñiga, 2001; Ruiz & Bahamonde, 2003; see also figs. 26 and 27) in addition to a few mountain lakes (Tartarotti et al., 1997; see figs. 26 and 27).

The species reported for these water bodies are mainly the calanoid, *Tumeodiaptomus diabolicus*, the cyclopoid *Mesocyclops longisetus*, and cladocerans such as *Daphnia pulex*, *Ceriodaphnia dubia*, and *Neobosmina chilensis* [= *Eubosmina hagmanni*] (cf. Schmid-Araya & Zúñiga, 1992; Ramos et al., 1998;

Fig. 27. Photographs of: top, Aculeo lagoon; centre, Rapel reservoir; and, bottom, Negra lagoon. [Photos P. R. De los Ríos-Escalante.]

TABLE III
Geographical characteristics and chlorophyll *a* concentration of some central Chilean lakes, lagoons, and reservoirs

Site	Surface (km^2)	Maximum depth (m)	Chl *a* (μg/L)	Number of species
Peñuelas lagoon	19.00	15	53.6	11
Rungue reservoir	0.48	15	65.87	4
Aculeo lagoon	12.00	5	102.4	4
El Plateado reservoir	0.02	13	No data	9
Rapel reservoir	13750.00	75	23.42	10
Lanalhue lake	32.00	26	5.2	5
Lleulleu lake	40.00	46	1.2	4

Muñoz et al., 2001; Ramos-Jiliberto & Zúñiga, 2001; Ruiz & Bahamonde, 2003; see table III). In this chapter, we shall consider a compilation of species richness, trophic status, and geographical features of six central Chilean lakes and reservoirs, and in order to compare those data, a PCA analysis was applied using the software Xlstat 5.0. The data fed into the analysis were obtained from the literature (Mühlhauser & Vila, 1987; Schmid-Araya & Zúñiga, 1992; Ramos et al., 1998; Muñoz et al., 2001; Ramos-Jiliberto & Zúñiga, 2001; Ruiz & Bahamonde, 2003) complemented by as yet unpublished observations (P. De los Ríos, unpubl. data; see tables III and IV). The results primarily reveal a significant, direct association between surface area and depth (table V, fig. 28). The results of the PCA revealed that the main regulating factors contribute 86.64%, the first axis explaining 25.24%, and the main variables are surface and maximum depth, whereas for the second axis the contribution was 61.40% with a marked role of chlorophyll *a* concentration in relation to the number of species found (fig. 28). The final results reveal that the Peñuelas and Aculeo lagoons have a high chlorophyll *a* concentration and a low species richness, whereas Runge reservoir and the Lanalhue and Lleulleu lakes show a low chlorophyll *a* concentration and a variable number of species; finally, Rapel reservoir has a high species number combined with a large surface area and a great maximum depth (fig. 28). Nonetheless, more information is needed to complete the overall picture, for example with regard to the large reservoirs (e.g., Colbun, and reservoirs located in the Coquimbo region), as well as for coastal lakes such as Vichuquen Lake and the Nahuelbutan Lakes. Only then we shall be able to compose a reliable overview of species' distributions all along the Chilean continental territory.

The results already exposed are, however, different in comparison to the published observations for Patagonian lakes and lagoons, where the low

TABLE IV

Species reported for some central Chilean lakes, lagoons, and reservoirs

Site	Species
Peñuelas lagoon	*Diaphanosoma chilense* Daday
	Ceriodaphnia dubia Richard
	Daphnia ambigua Scourfield
	Moina micrura Kurz
	Neobosmina chilensis (Daday) [= *Eubosmina hagmanni* (Stingelin)]
	Alona affinis Leydig
	Alona cambouei De Guerne & Richard [as: *A. pulchella* var. *cambouei*]
	Alona guttata G. O. Sars
	Chydorus sphaericus (O. F. Müller)
	Tumeodiaptomus diabolicus (Brehm)
	Tropocyclops prasinus meridionalis Kiefer
Rungue reservoir	*Moina micrura*
	Neobosmina chilensis [= *Eubosmina hagmanni*]
	Metacyclops mendocinus (Wierzejski)
	Tropocyclops prasinus meridionalis
Aculeo lagoon	*Neobosmina chilensis* [= *Eubosmina hagmanni*]
	Boeckella bergi Richard
	Mesocyclops longisetus (Thiébaud)
	Tropocyclops prasinus meridionalis
El Plateado reservoir	*Ceriodaphnia dubia*
	Daphnia ambigua
	Moina micrura
	Neobosmina chilensis [= *Eubosmina hagmanni*]
	Alona affinis
	Alona guttata
	Chydorus sphaericus
	Tumeodiaptomus diabolicus
	Mesocyclops longisetus
Rapel reservoir	*Diaphanosoma chilense*
	Ceriodaphnia dubia
	Daphnia ambigua
	Moina micrura
	Neobosmina chilensis [= *Eubosmina hagmanni*]
	Alona affinis
	Alona guttata
	Chydorus sphaericus
	Tumeodiaptomus diabolicus
	Mesocyclops longisetus

TABLE IV
(Continued)

Site	Species
Lanalhue lake	*Diaphanosoma chilense* *Ceriodaphnia dubia* *Neobosmina chilensis* [= *Eubosmina hagmanni*] *Tumeodiaptomus diabolicus* *Tropocyclops prasinus meridionalis*
Lleulleu lake	*Ceriodaphnia dubia* *Neobosmina chilensis* [= *Eubosmina hagmanni*] *Tumeodiaptomus diabolicus* *Tropocyclops prasinus meridionalis*

TABLE V
Correlation matrix and results of the PCA analysis for central Chilean lagoons and reservoirs; **bold** denotes a significant correlation

	Maximum depth	Chl *a*	Number of species
Surface area	**0.844**	−0.233	0.550
Maximum depth		−0.663	0.369
Chl *a*			−0.103

	F1	F2
Surface	31.875	7.900
Maximum depth	36.890	4.222
Chl *a*	15.620	49.605
Number of species	15.616	38.273

number of species is caused by the oligotrophy of those water bodies (Soto & Zúñiga, 1991; Soto & De los Ríos, 2006; De los Ríos & Soto, 2007). The results from these Patagonian waters also disagree with those reported from Northern Hemisphere lakes that show a direct relation between surface area and species number (Dodson, 1991, 1992; Dodson et al., 2000; Pinto-Coelho et al., 2005; Dodson & Silva-Briano, 2006). Nevertheless, the results observed for central Chilean lakes and reservoirs agree with the observations made for Northern Hemisphere lakes, where it is possible to find an increase in species richness at sites situated in surroundings where the environment has been altered due to human intervention (Dodson et al., 2005, 2007; Hoffman & Dodson, 2005).

From a biogeographical point of view, the crustacean zooplankton listed for central Chile does not comprise endemic species, as many of the species

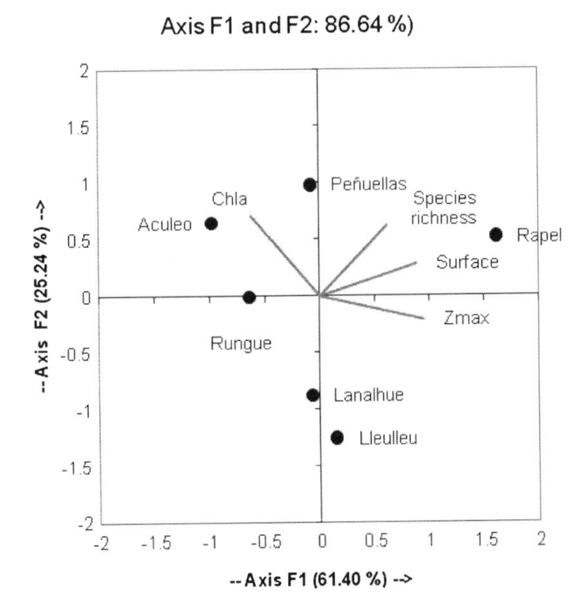

Fig. 28. Results of the PCA analysis of the sites studied in central Chile; see text for further explanation.

reported are widespread along the Chilean continental territory (Araya & Zúñiga, 1985; Menu-Marque et al., 2000). If we consider that calanoids are dominant along a wide latitudinal gradient in lakes and lagoons of continental Chile (Soto & Zúñiga, 1991), the main representative species is *Tumeodiaptomus diabolicus*. This species was reported from all lakes and reservoirs located in valleys at moderate altitudes (Schmid-Araya & Zúñiga, 1982; Ramos et al., 1998; Muñoz et al., 2001; Ramos-Jiliberto & Zúñiga, 2001; Ruiz & Bahamonde, 2003), whereas in mountain lakes *T. diabolicus* is regularly replaced by *Boeckella gracilipes* (cf. Cabrera et al., 1997). This observation would agree with the first descriptions of Zúñiga (1975), who also mentioned that *T. diabolicus* is dominant and practically exclusive in both coastal and relatively warm lakes, whereas in mountain waters and otherwise cold lakes this species is replaced by *B. gracilipes*. Similar results were reported from a comparative study of zooplankton assemblages of Chilean Patagonian lakes (De los Ríos & Soto, 2007). However, more ecological and experimental studies are obviously required to fully understand the potential species interactions that could explain this observation.

THE ECOLOGY OF THE CRUSTACEAN ZOOPLANKTON IN GLACIAL LAKES OF NORTHERN AND CENTRAL PATAGONIA

Chapter summary. — The northern and central parts of Patagonia have numerous large and deep lakes of glacial origin, which are originally oligotrophic and accommodate assemblages of crustacean plankton characterized by the dominance of calanoid copepods instead of a predominance of daphniid cladocerans, and by low species richness. However, due to current anthropogenic intervention, a transition from oligotrophy to mesotrophy has been observed, with, as a consequence, the predictable alterations in zooplankton assemblages, i.e., a specific increase in the abundance of daphniids and an increase in numbers of species. Another important factor that nowadays would affect these zooplankton assemblages, is their increased exposure to natural ultraviolet radiation due to the atmospheric ozone depletion at Antarctic and sub-Antarctic latitudes. The representative species here are calanoids such as *Boeckella gracilipes*, *B. gracilis*, *B. michaelseni*, and *Tumeodiaptomus diabolicus*, cladocerans such as *Ceriodaphnia dubia*, *Daphnia ambigua*, *D. pulex*, and *Neobosmina chilensis* [= *Eubosmina hagmanni*], and cyclopoids such as *Mesocyclops araucanus* and *Tropocyclops prasinus meridionalis*. The descriptions of Chilean lakes agree with those of Argentinean and New Zealand lakes, whereas they are markedly different in comparison to Northern Hemisphere lakes that have a dominance of cladocerans and higher numbers of species. Yet, the species richness of Chilean lakes is only in part similar to the patterns observed for Northern Hemisphere lakes, because no direct correlation with the surface has been reported, and only a weak, squared correlation with primary productivity was found.

Introduction

Northern Patagonia has many large, deep lakes located close to the Andes mountains that are of glacial origin (Thomasson, 1963; Campos, 1984; Soto & Zúñiga, 1991). The oligotrophy of these lakes is due to the original, native perennial forests in the catchment basins, from which hardly any nutrient input reached the lake (Soto et al., 1994; Soto & Campos, 1995; Steinhart et al.,

1999, 2002; Soto, 2002; Wölfl et al., 2003). However, during the last decades the native forest has for a large part been replaced by agricultural, industrial, and urban zones that have generated an increase in the input of nutrients from the surrounding catchment basin to the lake, with the inevitable transition of oligotrophy to mesotrophy (Soto, 2002; Wölfl et al., 2003). These alterations have brought about significant changes in the zooplankton assemblages, in particular evoking an increase in the abundance of daphniids (Villalobos, 1994, 2002; Villalobos & Geller, 1997; De los Ríos & Soto, 2007a).

The original crustacean zooplankton assemblages are characterized by an only sparse abundance of daphniids in comparison to calanoids, and by low species richness (Soto & Zúñiga, 1991; De los Ríos & Soto, 2006, 2007a, b; Soto & De los Ríos, 2006). The cause of this pattern is the oligotrophy, because daphniids have an only low tolerance to the oligotrophic status of a water body (Sterner & Hessen, 1994). Similar descriptions have been published for Argentinean Patagonian lakes, that are ultra-oligotrophic and completely unpolluted (Modenutti et al., 1998), as well as for some New Zealand lakes, that have a trophic status between oligotrophy and mesotrophy (Jeppensen et al., 1997, 2000). The representative species in these Patagonian deep lakes are, in this very order, the calanoids, *Boeckella gracilipes*, *Tumeodiaptomus diabolicus*, and *B. michaelseni* (cf. Araya & Zúñiga, 1985; Soto & Zúñiga, 1991; Bayly, 1992a, b; Menu-Marque et al., 2000; De los Ríos & Soto, 2007a). Also, the presence of cladocerans such as *Ceriodaphnia dubia*, *Daphnia ambigua*, *D. pulex*, *Neobosmina chilensis* [= *Eubosmina hagmanni*], and *Diaphanosoma chilense* is reported, and of cyclopoids like *Mesocyclops longisetus* and *Tropocyclops prasinus* (cf. Thomasson, 1963; Soto & Zúñiga, 1991; Soto & De los Ríos, 2006; De los Ríos & Soto, 2007b).

Other important sites located in northern and central Patagonia are small and deep lakes located in mountain zones with perennial native forest in their catchment basins, composed of *Araucaria araucana*, *Nothofagus pumilio*, *N. alpina* (Poeppig & Endler) Oerst., *N. dombeyi* (Mirb.) Blume, and *Fitzroya cupressoides* (cf. Steinhart et al., 1999, 2002; Luebert & Plitscoff, 2006). Due to this condition, these lakes are pristine, ultra-oligotrophic, and unpolluted, which data come from primary studies, both descriptive and experimental (Steinhart et al., 1999, 2002; De los Ríos et al., 2007a, b; De los Ríos & Romero-Mieres, 2009). The limnological properties of these ecosystems have been, and still are, only poorly studied due to notorious access problems, because many of these lakes can only be reached on foot over long and rough mountain paths (Steinhart et al., 1999, 2002; De los Ríos et al., 2007).

The zooplankton assemblages would comprise species such as *Boeckella gracilis*, *Daphnia pulex*, *Diaphanosoma chilense*, and *Neobosmina chilensis* [= *Eubosmina hagmanni*] (cf. Löffler, 1962; Thomasson, 1963; De los Ríos et al., 2008a, b, c; Soto & Zúñiga, 1991). Also, many of these studies have to be done in areas under protection of the Chilean government, or in private protected areas, because those are most important from a conservational point of view, due to their unpolluted and truly pristine condition (Steinhart et al., 1999, 2002; De los Ríos, 2003; De los Ríos et al., 2007; De los Ríos & Romero-Mieres, 2009).

Geographical features

The northern and central parts of Patagonia (38-51°S; see fig. 29) are characterized by the rainy climate and the perennial forest that made the original environment in the catchment basins of the water bodies located there (Thomasson, 1963; Niemeyer & Cereceda, 1984; Luebert & Plitscoff, 2006). These forests have largely been replaced during the last decades by agricultural, industrial, and urban zones (Soto & Campos, 1995; Soto, 2002; De los Ríos & Soto, 2007a, b). At 37-39°S the vegetation communities are characterized by *Nothofagus pumilio*, *N. alpina*, and *N. dombeyi* at altitudes lower than 1000 m a.s.l., whereas at higher altitudes *Araucaria araucana* is dominant (Luebert & Plitscoff, 2006; De los Ríos et al., 2007; De los Ríos

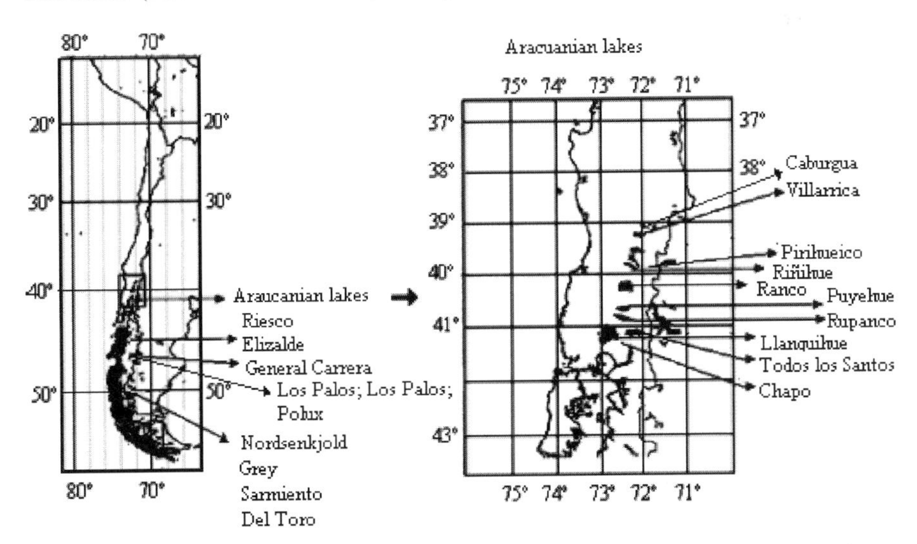

Fig. 29. Map of the sites in northern and central Patagonia included in the present study.

& Romero-Mieres, 2009). South of 39°S (fig. 30) *A. araucana* disappears, and is gradually replaced by *Fitzroya cupressoides* (cf. Steinhart et al., 1999; Luebert & Plitscoff, 2006). Finally, south of 42°S, the dominant species in the native forests is *Nothofagus antarctica* (cf. Luebert & Plitscoff, 2006) (figs. 31, 32). This zone is characterized by its glacial origin, that has generated the characteristic geomorphology of the lakes and rivers (Campos, 1984; Pedrozo et al., 1993; Timms, 1995). Some of the lakes still have glaciers associated with them, such as Todos los Santos Lake (41°S; Campos et al., 1990) as well as some of the large lakes in Torres del Paine National Park (Campos et al., 1994a; Soto et al., 1994) (fig. 32).

From a geomorphological point of view, the northern Patagonian lakes are deep and show a so-called "cryptodepression" [= a part that lies below sea level] (Campos, 1984; Soto & Stockner, 1996), with a marked slope and either with an east-west direction of their longer axis, or else with a sub-circular shape, all due to the former glacier activities (Campos, 1984; Soto & Campos, 1995; Soto & Stockner, 1996). When the lakes have associated glaciers, these have a characteristic colour, such as green in Todos los Santos Lake (Campos et al., 1990), and blue-green or white in the lakes of Torres del Paine National Park (Soto et al., 1994). Nevertheless, the waters of many of the Chilean and also Argentinean Patagonian lakes are very transparent, due to their low mineral contents, oligotrophy, and low concentrations of dissolved organic carbon (Morris et al., 1995; Marinone et al., 2006).

From a limnological perspective, the Chilean Patagonian lakes are warm-monomictic, with only one mixing period, in winter, and a stratification period in spring and summer (Campos, 1984). An important characteristic is the deep mixing layer or epilimnion, which increases at southern latitudes due to the exposition to the strong winds (Geller, 1992; Soto, 2002). The thermocline can reach a depth of 75 m at 51°S (Campos et al., 1994a, b). The mixing depth is an important regulator of trophic status, because light limitation affects the photosynthetic activity of the phytoplankton, and low light exposure periods occur at greater depths (Sterner et al., 1997; Soto, 2002). These conditions are markedly different in comparison to Northern Hemisphere lakes, where the chlorophyll *a* concentration is directly related to the total phosphorus concentration and a low value of the total phosphorus/total nitrogen molar ratio (Guildford & Hecky, 2000). Therefore, a direct correlation of phosphorus concentration with daphniid abundance has been described for those ecosystems (Elser et al., 1996, 2000).

Fig. 30. Photographs of: top, Lake Conguillío, located in a protected zone with a native forest of *Araucaria araucana* (Molina) K. Koch; and, bottom, Villarrica Lake, with the homonymous volcano in the background. [Photos P. R. De los Ríos-Escalante.]

Fig. 31. Photographs of: top, Lake Llanquihue with the Osorno volcano; and, bottom, Lake Sargazo: this lake is located in a protected area with native forests of *Fitzroya cupressoides* (Molina) Johnston. [Photos P. R. De los Ríos-Escalante.]

Fig. 32. Photographs of some lakes in Torres del Paine National Park (both top and bottom), at sites where the presence of icefields and steep mountain ranges is particularly evident. [Photos P. R. De los Ríos-Escalante.]

Species assemblages

On the basis of the available literature (Campos et al., 1982, 1983, 1987a, b, 1988, 1990, 1992a, b, 1994a, b; Soto et al., 1994; Wölfl, 1996; Villalobos, 1999; Villalobos et al., 2003a; see table VI), a PCA analysis was performed. This analysis revealed an inverse relationship between species number and latitude, and also with mixing depth (table VII). Also the two factors, maximum depth and area, were the most important variables (table VII). A non-parametric correlation analysis was applied as well, and this demonstrated the

TABLE VI

Latitude, maximum depth, mixing depth, surface area, chlorophyll *a* concentration, primary productivity, and species richness at the sites included in the study of the lakes of northern and central Patagonia

	Lat. S, Long. W	Z_{max} (m)	Z_{mix} (m)	Area (km^2)	Chl *a* (μg/L)	Primary productivity ($mgC/m^3/h$)	Species richness	Reference
Caburgua	39°07′ 71°45′	327	15	51.9	0.1	No data	7	(1)
Villarrica	39°18′ 72°07′	185	15	175.8	0.4	8.20	7	(2)
Riñihue	39°50′ 72°20′	323	20	77.5	1.2	5.10	7	(3)
Pirihueico	39°50′ 71°48′	145	20	30.5	0.6	0.42	5	(3)
Ranco	40°13′ 70°22′	199	20	442.6	0.8	3.90	9	(4)
Puyehue	40°40′ 72°30′	123	No data	165.4	2.1	1.40	7	(5)
Rupanco	40°50′ 72°30′	273	No data	235.8	1.2	1.55	4	(6)
Llanquihue	41°08′ 72°50′	317	40	870.5	0.5	1.90	3	(7)
Todos los Santos	41°08′ 72°50′	335	40	178.5	0.4	0.40	4	(8)
Chapo	41°27′ 72°30′	298	No data	45.3	0.3	1.95	6	(9)
Los Palos	45°19′ 72°42′	59	No data	5	No data	No data	4	(10)
Riesco	45°39′ 72°20′	130	No data	14.7	0.9	No data	4	(10)
Pólux	45°43′ 71°53′	10	No data	9	No data	No data	4	(11)
Elizalde	45°44′ 72°25′	130	No data	30	No data	No data	3	(11)
Escondida	45°49′ 72°40′	43	No data	7	No data	No data	5	(10)
General Carrera	45°50′ 72°00′	410	No data	1892	No data	No data	3	(12)
Sarmiento	51°03′ 72°37′	312	75	86	0.3	No data	5	(13)
Grey	51°03′ 72°53′	200	75	15	0.4	No data	2	(13)
Nordsdenkjold	51°03′ 72°56′	200	75	25	0.3	No data	2	(13)
Del Toro	51°12′ 72°38′	317	75	196	0.4	No data	4	(13)

Legends: Z_{max}, maximum depth; Z_{mix}, mixing depth; Chl *a*: chlorophyll *a* concentration. References: (1) Campos et al. (1987b); (2) Campos et al. (1983); (3) Wölfl (1996); (4) Campos et al. (1992a); (5) Campos et al. (1989); (6) Campos et al. (1992b); (7) Campos et al. (1988); (8) Campos et al. (1990); (9) Villalobos et al. (2003); (10) Villalobos (1999); (11) De los Ríos (in press); (12) De los Ríos & Soto (2007); (13) Soto & De los Ríos (2006).

TABLE VII

Correlation matrix of parameters considered in the study of northern and central Patagonian lakes; values in **bold** denote significant correlations ($p < 0.05$), and percentages of importance for the factors studied in the Chilean lakes here reported upon

	Z_{max} (m)	Z_{mix} (m)	Surface area (km^2)	Chl *a* (μg/L)	Species richness
Latitude	−0.109	**0.939**	−0.304	−0.307	**−0.643**
Z_{max}		0.130	0.361	−0.321	0.074
Z_{mix}			−0.124	−0.399	**−0.681**
Area				0.050	0.072
Chl *a*					0.259

Percentage of importance of PCA factors

	F1	F2
Latitude	31.937	1.626
Z_{max}	0.009	51.763
Z_{mix}	32.738	0.754
Area	2.719	30.424
Chl *a*	9.411	15.353
Species richness	23.187	0.080

Legend: Z_{max}, maximum depth; Z_{mix}, mixing depth; Chl *a*, chlorophyll *a* concentration.

existence of inverse correlations between species richness and latitude, and mixing depth, respectively, whereas no significant relationships with maximum depth, area, or chlorophyll *a* concentrations were found (fig. 33). The PCA results show two main groups, viz., oligo-mesotrophic lakes with high species richness, and oligotrophic lakes with low species richness (fig. 34). Finally, a weak, squared correlation between primary productivity and number of species could be demonstrated (fig. 35).

The crustacean zooplankton reported for Patagonian lakes is characterized by low species richness (Zúñiga & Zúñiga, 1977, 1978; Soto & Zúñiga, 1991; Soto & De los Ríos, 2006; De los Ríos & Soto, 2007b). The species assemblages are characterized by the presence of calanoids such as *Boeckella gracilipes*, that is distributed between 38 and 44°S and coexists with *Tumeodiaptomus diabolicus* at 38-42°S (Zúñiga, 1975; Villalobos et al., 2003b), while south of 44°S *Boeckella michaelseni* is dominant (Soto & Zúñiga, 1991; Menu-Marque et al., 2000; De los Ríos & Soto, 2007a). In high mountain lakes at latitudes of 38-40°S, *Boeckella gracilis* is a representative species (Araya & Zúñiga, 1985; De los Ríos et al., 2007, 2009; De los Ríos & Romero-Mieres, 2009). Also, cladocerans such as *Ceriodaphnia dubia*, *Daphnia ambigua*, *D. pulex*, and *Diaphanosoma chilense* appear between 38 and 44°S, while *Neo-*

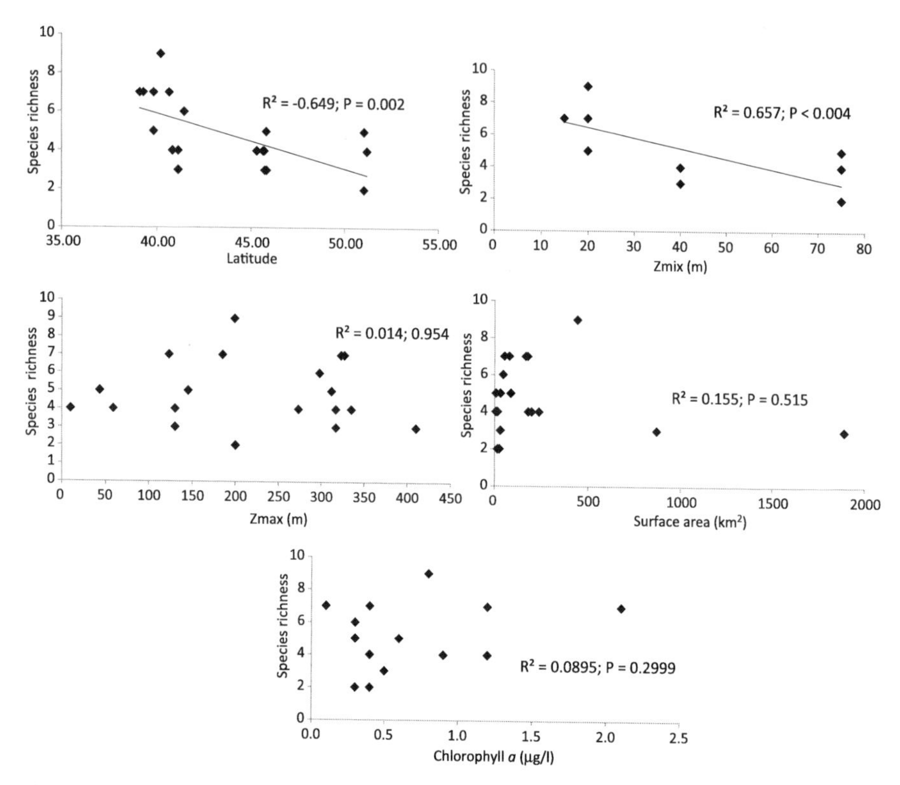

Fig. 33. Graphs of correlations between species richness of crustacean zooplankton (y-axis) and the parameters of the lakes (x-axis) in northern and central Patagonia considered in the present study; Z_{max}, maximum depth; Z_{mix}, mixing depth.

bosmina chilensis [= *Eubosmina hagmanni*] is widespread (Deevey & Deevey, 1971; Soto & Zúñiga, 1991; De los Ríos & Soto, 2007b; De los Ríos, 2008). The cyclopoids seem to be similary distributed along 38-51°S, and the most representative species would be *Mesocyclops longisetus* and *Tropocyclops prasinus* (cf. Thomasson, 1963; Soto & Zúñiga, 1979; Soto & De los Ríos, 2006; De los Ríos & Soto, 2007b). According to the literature, oligotrophy would be the main factor that causes the low species richness in Chilean lakes (Soto & Zúñiga, 1991; Soto & De los Ríos, 2006; De los Ríos & Soto, 2007). A similar pattern was also found in Argentinean Patagonian lakes (Modenutti et al., 1998; Quiros & Drago, 1999).

Another important characteristic is the marked calanoid predominance with respect to daphniid cladocerans. In contrast, in North American lakes the daphniids are dominant (Soto & Zúñiga, 1991; Gillooly & Dodson, 2000). This pattern of calanoid dominance is also seen in Argentinean lakes (Modenutti et al., 1998) as well as in New Zealand (Jeppensen et al., 1997, 2000). The

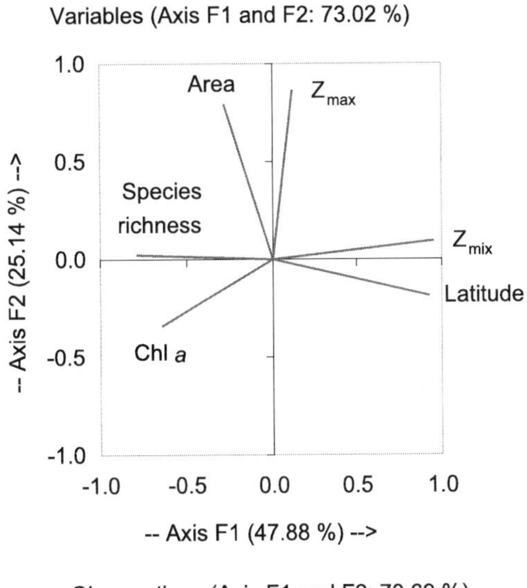

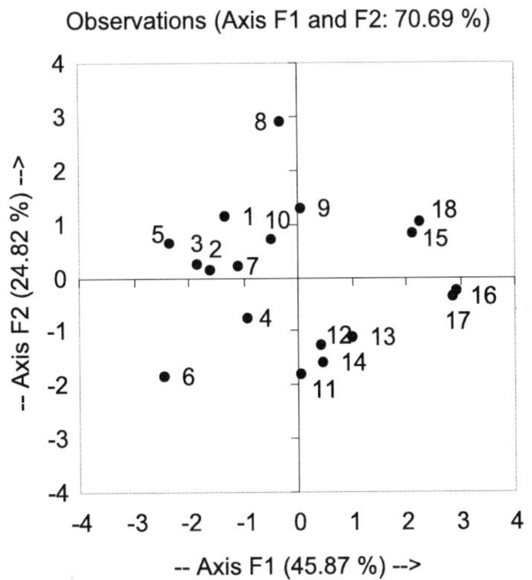

Fig. 34. Principal component analysis of the sites studied in northern and central Patagonia. For explanation, see text. Legends: Z_{max}, maximum depth (m); Z_{mix}, mixing depth (m); Chl *a*: chlorophyll *a* concentration (mg/L). Localities: 1, Caburgua; 2, Villarrica; 3, Riñihue; 4, Pirihueico; 5, Ranco; 6, Puyehue; 7, Rupanco; 8, Llanquihue; 9, Todos los Santos; 10, Chapo; 11, Los Palos; 12, Riesco; 13, Pólux; 14, Elizalde; 15, Escondida; 16, General Carrera; 17, Sarmiento; 18, Grey; 19, Norsdenkjold; 20, Del Toro.

classical literature on zooplankton ecology described a high predominance of daphniids in zooplankton assemblages (Sterner & Hessen, 1994; Gillooly & Dodson, 2000), correlated with high total phosphorus concentrations and low values of molar ratios of total phosphorus and total nitrogen, a value which it is

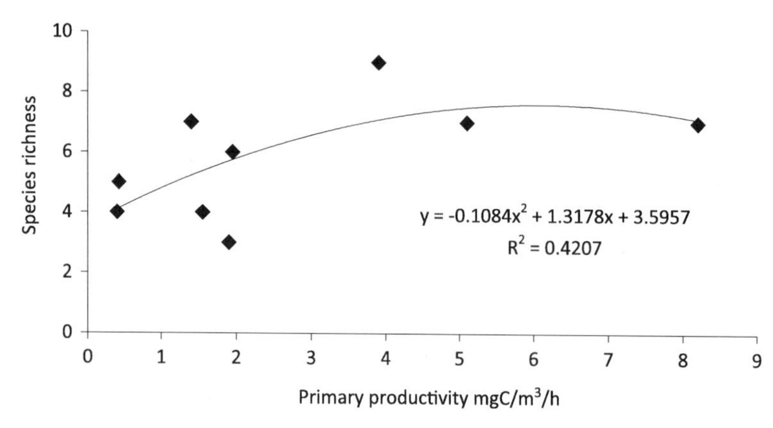

Fig. 35. Relation between primary productivity and species richness for the sites studied in northern and central Patagonia.

directly related to chlorophyll *a* concentration (Elser et al., 1996, 2000). As a consequence, a direct correlation between daphniid abundance and chlorophyll *a* concentration was found (Sterner & Hessen, 1994), and both are related to total phosphorus concentration, i.e., low values of total phosphorus, and inversely related to the molar ratio total nitrogen/total phosphorus (Guildford & Hecky, 2000). This was not found in Chilean lakes, because, although these have low values of the total nitrogen/total phosphorus molar ratio, they also show low absolute concentrations of both nutrients (Steinhart et al., 1999, 2002) that would not sustain abundant daphniid populations (Soto & Stockner, 1996). These results would explain the direct relationship of daphniid abundance and chlorophyll *a* concentrations observed in Chilean lakes (De los Ríos & Soto, 2006; Soto & De los Ríos, 2006), whereas there was no correlation with the nitrogen/phosphorus molar ratio (Soto & Stockner, 1996; De los Ríos, 2003). In conclusion, the zooplankton assemblages found in these Chilean water bodies seem not to be regulated by mechanisms similar to those in Northern Hemisphere lakes (De los Ríos, 2003; De los Ríos & Soto, 2006).

Calanoid dominance: comparison with other, similar cases

The dominance of calanoids was already observed during the first studies of Patagonian lakes (Thomasson, 1963; Domínguez & Zúñiga, 1979; Campos, 1984; Soto & Zúñiga, 1991). The first ecological study that proposed that oligotrophy would be the main cause of calanoid dominance and low daphniid relative abundance (Soto & Zúñiga, 1991) was confirmed by De los Ríos & Soto (2006) and Soto & De los Ríos (2006). These results agree with those from Argentinean Patagonian lakes (Modenutti et al., 1998; Quiros & Drago, 1998) and from lakes in New Zealand (Jeppensen et al., 1997, 2000). The studies denoted that in oligotrophic Patagonian lakes, mainly between 38 and 41°S, there is a dominance of mixotrophic ciliates such as *Stentor araucanus* Foissner & Wolel, 1994 and *S. amethystinus* Leidy, 1880 (cf. Wölfl, 1996; Wölfl & Geller, 2002; Wölfl, 2007) or of *Ophyridium* species (cf. Modenutti & Pérez, 2003). In the trophic web, these mixotrophic ciliates are the main preys for calanoid copepods (Wölfl, 1996, 2007; Balseiro et al., 2001; Kamjunke et al., 2009). A different situation occurs in southern Patagonian lakes, where diatoms are dominant in phytoplankton assemblages (Campos et al., 1994a, b), due to the great mixing depth. This depth can increase from 50 to 75 meters (Geller, 1992; Soto, 2002), and only allows the presence of diatoms, which are able to tolerate this physical stress (Wölfl, 1995). Different conditions are found at northern Patagonian sites where, due to transitions from oligotrophy to mesotrophy, mixotrophic ciliates and diatoms are replaced by planktonic Chlorophyta (cf. Villalobos & Geller, 1997; Villalobos, 2002; Villalobos et al., 2003), with as a consequence an increased abundance of daphniids (Villalobos & Geller, 1997; Villalobos, 1994, 2002). Thus, calanoids would play a key role as grazers, and on their turn as prey for zooplanktivorous fish (Soto & Zúñiga, 1991; De los Ríos, 2003; Soto & De los Ríos, 2006). This is because, fishes are visual predators (Winder, 2003), calanoids are dominant in oligotrophy, and the Chilean lakes are transparent as a result of that very oligotrophic condition — and in addition, in many situations the calanoids are larger than the cladocerans (Araya & Zúñiga, 1985; De los Ríos, 2003). In these transparent water bodies, the fishes would thus predate primarily on the calanoid copepods as these comprise the bigger preys that are easier to spot visually (Soto & Zúñiga, 1991; Soto et al., 1994; Modenutti et al., 1998, 2003; De los Ríos, 2003). Unfortunately, these trophic zooplankton-fish interactions have not been studied, while there is also an unknown effect due to the presence of introduced salmonids (Soto & Zúñiga, 1991). In their adult stages these predate on macroinvertebrates (Soto & Campos, 1995), but it is probable that their juvenile stages will predate on zooplankton (Wölfl, 2007). A similar pattern is expected in high mountain lakes, because these are oligotrophic as well (Steinhart et al., 1999, 2002) and also show calanoid dominance (De los Ríos et al., 2009). The presence of fish varies, because there are waters with salmonids, while others have only native fishes such as *Galaxias* spp. and the water bodies at issue are either permanent with a moderate depth, or shallow and ephemeral (De los Ríos et al., 2007, 2008). In these small lakes, ponds, and pools, the dominant species is *Boeckella gracilis*, that is large in comparison with *B. gracilipes* and *Tumeodiaptomus diabolicus* (cf. Araya & Zúñiga, 1985). This species is also pigmented, which is thought to be an adaptation, probably as a potential protective response against exposure to natural ultraviolet radiation (Tartarotti et al., 2004).

Zooplankton assemblages and a comparison with North American counterparts

The first comparative studies on zooplankton assemblages in Chilean lakes and preliminary comparisons with their North American counterparts were the descriptions of Soto & Zúñiga (1991) and Soto & Stockner (1996). The aim of both studies was to investigate if Chilean lakes are markedly different from the Northern Hemisphere lakes, on which the classical limnological descriptions had been based. The main difference reported was the dominance of calanoids over daphniids, largely due to oligotrophy (Soto & Zúñiga, 1991; Soto & Stockner, 1996; Gillooly & Dodson, 2000; Soto & De los Ríos, 2006; De los Ríos & Soto, 2006). Another important difference is the low species richness reported for Chilean lakes in comparison to the lakes of North America (De los Ríos & Soto, 2007b). The potential cause is again the oligotrophy of Chilean lakes, because, in a community-ecological sense, a site of low productivity has a low species richness, which, of course, also holds true for the realm of zooplankton ecology (Modenutti et al., 1998; Soto & De los Ríos, 2006). The causes of these differences are the various forms of human intervention that generate ecological changes in the environment, in particular an increase in trophic level allows a habitat to sustain a higher species richness than in the original, oligotrophic conditions used to be possible (Dodson et al., 2005, 2007; Hoffman & Dodson, 2005). In a similar sense, there also is a direct relation between the surface area of the water body and the species richness of the zooplankton community, again calculated on the basis of Northern Hemisphere lakes (Dodson, 1991, 1992; Dodson & Silva-Briano, 2006). However, this relationship was not observed for Chilean Patagonian lakes (Soto & Zúñiga, 1991; Soto & De los Ríos, 2006; De los Ríos & Soto, 2007b). Another important observation reported for zooplankton communities in Northern Hemisphere lakes, is the presence of a squared relationship between species number and primary productivity (Dodson et al.,

2000; Pinto-Coelho et al., 2005), with a maximum species richness reached at a primary productivity of 67-149 mgC/m^2/year (Dodson et al., 2000). The primary productivity reported for Chilean lakes is lower than 70 mgC/m^2/year (Montecino, 1991), and a weak squared relationship with species number may possibly be established (fig. 35). As yet, there are no detailed studies on either the primary productivity or the zooplankton ecology in Chilean lakes, however, at least not at any appreciable scale. This means that more studies will be necessary to describe the Chilean crustacean plankton populations and the communities to which they belong, in order to make fruitful comparisons with the available literature data pertaining to the Northern Hemisphere.

The potential role of natural ultraviolet radiation

Another important factor is the exposure to natural ultraviolet radiation, which was always there, but which has lately been increased due to the atmospheric ozone depletion at polar and subpolar latitudes (Morris et al., 1995; Villafañe et al., 2001; Díaz et al., 2006; Marinone et al., 2006). This phenomenon affects the Araucania region (De los Ríos et al., 2007a, b; see also table VIII, based on data of the Chilean Meteorological Region). Ultraviolet radiation penetrates into the water column due to the transparency of the water (Lowengreen et al., 1994; Morris et al., 1995; Laurion et al., 2000). The zooplankton organisms living there may develop a behaviour involving daily vertical migrations, as a photoprotective response, i.e., as a negative phototaxis that makes them avoid the euphotic zones where natural ultraviolet radiation penetrates (Storz & Paul, 1998; Rhode et al., 2000; Alonso et al., 2003; Winder, 2003). In this scenario, it is probable that the exposure to natural ultraviolet radiation would generate a selective force that influences the evolutionary pathways of the species constituting the local zooplankton assemblages (cf. intraspecific evolution) in which only the most evasive individuals will contribute to the next generation. However, the composition of those assemblages may be influenced just as well (cf. interspecific evolution), since only those species of such an assemblage will survive that are either the most tolerant to UV radiation, or have found the best solutions to cope with it, e.g., by becoming evasive, or pigmented, etc. (Marinone et al., 2006). Although, obviously, depletion of the community by species becoming extinct will necessarily change the existing ecosystem, in particular the food web, a new ecological balance will become established in time. This whole sequence would explain the observations made, because some taxa such as daphniids

TABLE VIII
UV-B radiation, daily maximum (W/m^2) and doses (kJ/m^2) for Temuco (38°41′S 72°35′W) (cf. De los Ríos et al., 2007)

Date	Daily maximum UV-B (W/m^2)	Doses (kJ/m^2)
28 Dec 2005	4.3	107.6
29 Dec 2006	3.5	86.8
03 Jan 2006	3.4	73.8
10 Jan 2006	4.0	91.4
25 Jan 2006	4.2	100.8
26 Jan 2006	4.2	97.6
01 Feb 2006	4.1	99.4
13 Feb 2006	4.2	101.9
02 Mar 2006	3.0	69.5
03 Mar 2006	3.2	59.4
17 Mar 2006	2.5	46.4
24 Mar 2006	2.0	43.2

Legend: W/m^2 = $Watt/m^2$.

are less tolerant to the exposure to natural ultraviolet radiation in oligotrophic environments (De los Ríos & Soto, 2005). That fact would explain, in its turn, the low species richness and the calanoid dominance in the ultra-oligotrophic and oligotrophic lakes of Patagonia, both in Argentina and in Chile (Marinone et al., 2006; De los Ríos & Soto, 2006, 2007). A different situation can be found in lakes with a transition from oligotrophy to mesotrophy, because here the daphniids are dominant (De los Ríos & Soto, 2006, 2007b). This is due to the availability of energy resources as well as to the depth of the habitats, which provides an opportunity to find protection against UV radiation, i.e., when moving out of the euphotic zone to greater depths (Villafañe et al., 2001; De los Ríos & Soto, 2006; Marinone et al., 2006). Similar conditions have been described for New Zealand lakes, where the exposure to natural ultraviolet radiation and the gradient of trophic status may also explain the observed differences in dominance of cladocerans (Vareshi & Wübbens, 2001).

The condition of (increased) exposure to natural UV radiation is aggravated due to the low concentrations of dissolved organic carbon (DOC) (Morris et al., 1995; Soto & Campos, 1995; Soto & Stockner, 1996; De los Ríos & Soto, 2006; De los Ríos et al., 2007). The humic substances that are measured as dissolved organic carbon, make a natural protection against the ultraviolet radiation that forms part of the sunlight (Morris et al., 1995) and thus generates the required photoprotection for the vulnerable zooplankton species (De los

Ríos, 2003, 2004). This low concentration of dissolved organic carbon is due to the natural vegetation of the catchment basin surrounding the water body, that absorbs the nutrients in the soil and hence impedes the input of those substances into the lake (Soto & Campos, 1995; Soto & Stockner, 1996; Steinhart et al., 1999, 2002). This condition has actually been observed in the field, i.e., in lakes in the Huerquehue National Park, where a direct relationship was found between humic substance concentrations and species richness (De los Ríos et al., 2007).

THE ECOLOGY OF THE CRUSTACEAN ZOOPLANKTON IN CENTRAL AND SOUTHERN PATAGONIAN SHALLOW PONDS

Chapter summary. — The central and southern Patagonian plains (44-54°S) have numerous shallow pools and lagoons, both permanent and ephemeral, without fish, that make nesting and feeding areas for aquatic birds such as ducks, swans, and flamingoes. The studies on aquatic invertebrates hitherto performed only describe the planktonic crustaceans, but these data already indicate these water bodies to harbour zooplankton assemblages characterized by their high species richness. Some of those species are endemic for southern Patagonia and the sub-Antarctic islands, and there are also widespread species. Calanoid copepods constitute the main dominant group in subsaline water bodies (<12 g/L), whereas in more saline waters the anostracan *Artemia persimilis* dominates. In some cases, this last-mentioned species can coexist with halophilic copepods, a situation obviously different from that encountered in northern Chilean saline lakes. The main regulating factors of zooplankton assemblages are salinity and trophic status, although it is possible that the exposure to natural ultraviolet radiation might also play a role of some importance as a regulator of species assemblages. Viewed in an ecological sense, the ecology of these water bodies is similar to that found in northern Chilean inland waters.

Introduction

The southern Chilean part of the region of Patagonia is characterized by its irregular geography due to the presence of numerous islands, inner seas, straits, ice fields, lakes, mountains, and rivers, that join to create a landscape interesting for studies, but at the same time render those studies difficult mainly as a result of the severe problems encountered in attempts at accessing the study sites (Niemeyer & Cereceda, 1984; Soto et al., 1994; De los Ríos, 2005; Soto & De los Ríos, 2006). From a limnological point of view there are many large, deep lakes (see previous chapter) and shallow, temporal and even ephemeral water bodies, which last group will be treated in the present chapter.

The shallow water bodies in central and southern Patagonia were perhaps the first water bodies studied in Chile (see chapter Introduction). However, these sites have been studied again in detail, practically eighty years after the first descriptions had been made (Soto et al., 1994), and those recent studies served as a basis for the recently performed ecological surveys (De los Ríos, 2005; De los Ríos & Contreras, 2005; De los Ríos et al., 2007a, b, 2008a, b, c; De los Ríos, 2008; De los Ríos & Rivera, 2008). These sites are similar to their counterparts in southern Argentinean Patagonia, and area that has been exhaustively studied already (Modenutti et al., 1998), as well as to comparable locations on the sub-Antarctic islands (Pugh et al., 2002; Dartnall et al., 2005). Yet, due to the access problems mentioned for southern Chilean Patagonia, studies are primarily restricted to Torres del Paine National Park (51°S) and the coastal zones of Magellan Strait (De los Ríos et al., 2008a, b).

Geographical characterization

Both from a geographical and an ecological point of view, this zone is located in a region with marked glacial influence. Some of the water bodies are associated with ice fields without appreciable physical barriers, which results in the condition that many waters are similar to their counterparts in the adjoining Argentinean plains (Niemeyer & Cereceda, 1984). The landscape has many large, extensive plains, because the Andes mountains are absent at those southernmost latitudes. The climate is subpolar, with rains in winter and strong winds in spring and summer (Soto et al., 1994; Luebert & Pliscoff, 2006). This last condition generates an increase in evaporation that would explain the high mineral contents in some of the water bodies found here (Soto et al., 1994; Campos et al., 1996; see table IX). Due the subpolar weather conditions and the harsh wind exposure, the vegetation of the catchment basins surrounding the water bodies is characterized by the presence of small forests of *Nothofagus antarctica*, and plains with shrubs and grasses such as *Festuca*. The subpolar climate also implies that sometimes the shallower water bodies are frozen in winter (Soto et al., 1994), and the exposure to the strong winds causes drying-out of the shallow water bodies in summer (De los Ríos et al., 2008a, b, c). The latter condition pertains, e.g., to some sites in Torres del Paine National Park (De los Ríos, 2008; De los Ríos et al., 2008c; see fig. 36), and on the island of Tierra del Fuego (De los Ríos, 2005; fig. 37), as well as to some sites close to Balmaceda airport (45°S; De los Ríos, 2008). Also, many of these waters are located rather close to each other, which creates marked

TABLE IX

Geographical location, maximum depth (Z_{max}), surface, conductivity, and number of species reported for southern Patagonian shallow lagoons

Site	Sampling date	Location	Z_{max} [m]	Surface [km^2]	Conductivity [μS/cm]	Number of species	Reference
1-Isidoro	October 2001	50°57′ 72°53′	<1.5	<0.1	150.00	6	De los Ríos (2005)
2-Guanaco	October 2001	51°01′ 72°50′	<2	<0.1	552.00	7	De los Ríos (2005)
3-Don Alvaro	October 2001	51°01′ 72°52′	<2	<0.1	153.00	7	De los Ríos (2005)
4-Larga	1989-1991	51°01′ 72°52′	<5	<0.1	3448.00	6	Soto & De los Ríos (2006)
5-	March 2001				4330.00	4	De los Ríos (unpubl. data)
6-	October 2001				20300.00	5	De los Ríos (unpubl. data)
7-Redonda	1989-1991	51°01′ 72°52′	<3	<0.1	1486.00	8	Soto & De los Ríos (2006)
8-	March 2001				1365.00	4	De los Ríos (unpubl. data)
9-	October 2001				643.00	6	De los Ríos (unpubl. data)
10-Juncos	1989-1991	51°01′ 72°52′	<3	<0.1	2288.00	6	Soto & De los Ríos (2006)
11-	March 2001				403.20	4	De los Ríos (unpubl. data)
12-	October 2001				503.50	5	De los Ríos (unpubl. data)
13-Cisnes	1989-1991	51°01′ 72°52′	1.0	<0.1	16560.00	7	Soto & De los Ríos (2006)
14-	March 2001				13200.20	4	De los Ríos (unpubl. data)
15-	October 2001				5900.00	5	De los Ríos (unpubl. data)
16-Jovito	1989-1991	51°02′ 72°54′	<3	<0.1	1380.00	6	Soto & De los Ríos (2006)
17-Paso de la Muerte	1989-1991	51°02′ 72°55′	<3	<0.1	801.00	7	Soto & De los Ríos (2006)
18-Paso de la Muerte	March 2001				498.0	6	De los Ríos (unpubl. data)
19-Paso de la Muerte	October 2001				563.0	3	De los Ríos (unpubl. data)
20-Vega del Toro	October 2001	51°07′ 71°40′	<1.5	<0.1	2156.00	8	De los Ríos (2005)
21-Monserrat	October 2001	51°07′ 72°47′	<1.5	<0.1	362.40	7	De los Ríos (2005)
22-Kon Aikén	October 2001	52°50′ 71°40′	<1.0	<0.1	962.00	7	De los Ríos (2005)
23-Kon-Aikén 1	October 2005	52°51′ 70°55′	<1.0	<0.1	640.00	4	De los Ríos et al. (2008a, b)
24-Kon-Aikén 2	October 2006	52°51′ 70°55′	<1.0	<0.1	700.00	4	De los Ríos et al. (2008a, b)
25-Kon-Aikén 3	October 2006	52°51′ 70°55′	<1.0	<0.1	480.00	4	De los Ríos et al. (2008a, b)
26-Kon-Aikén 4	October 2006	52°51′ 70°55′	<1.0	<0.1	420.00	4	De los Ríos et al. (2008a, b)
27-Kon-Aikén 5	October 2006	52°51′ 70°55′	<1.0	<0.1	450.00	4	De los Ríos et al. (2008a, b)
28-Kon-Aikén 6	October 2006	52°51′ 70°55′	<1.0	<0.1	560.00	4	De los Ríos et al. (2008a, b)
29-Laredo 1	October 2006	52°51′ 70°55′	<2.0	<0.1	6.140	4	De los Ríos et al. (2008a, b)
30-Laredo 2	October 2006	52°51′ 70°55′	<2.0	<0.1	2.290	3	De los Ríos et al. (2008a, b)
31-Rio Seco	October 2006	53°06′ 70°53′	<1.0	<0.1	2.300	3	De los Ríos et al. (2008a, b)

Fig. 36. Photographs of some shallow lagoons in the Magellan region: on the left, Amarga Lagoon in Torres del Paine National Park; to the right, Ana Lagoon in Pali Aike National Park. [Photos P. R. De los Ríos-Escalante.]

Fig. 37. Photographs of two different shallow lagoons on Tierra del Fuego Island. [Photos P. R. De los Ríos-Escalante.]

similarities, in particular with regard to having various species in common, due to the easy dispersal, obviously facilitated by the short distances involved (Menu-Marque et al., 2000).

Species assemblages

The studies that have been performed describe a relatively high species richness in these water bodies (Soto & De los Ríos, 2006; De los Ríos, 2008; De los Ríos et al., 2008a, b, c; table IX). At low conductivity, the representative species are *Branchinecta gaini*, *B. granulosa*, *B. vuriloche*, *Daphnia dadayana*, *Boeckella brasiliensis*, *B. gracilipes*, *B. michaelseni*, *B. poppei*, and *Parabroteas sarsi* (cf. Soto & De los Ríos, 2008; De los Ríos et al., 2008a, b, c), whereas at moderate and high conductivity the anostracan *Artemia persimilis* is present (Gajardo et al., 1998; De los Ríos & Zúñiga, 2000; De los Ríos, 2005). Also, the literature describes the presence of widespread species such as the calanoids *Boeckella gracilipes* and *B. poopoensis* (cf. Menu-Marque et al., 2000), as well as cladocerans such as *Daphnia ambigua*, *D. pulex*, *Chydorus sphaericus*, and *Neobosmina chilensis* [= *Eubosmina hagmanni*] (cf. Soto & De los Ríos, 2006; De los Ríos, 2008; De los Ríos et al., 2008a, b, c), that can all tolerate low to moderate conductivity values (Paggi, 1996a). Another important characteristic is the presence of endemic calanoids that have been described only for southern Patagonia and/or some sub-Antarctic islands, such as *Boeckella brevicaudata*, that has been described only once from continental Chilean territory (De los Ríos, 2005, 2008). Other species have not (yet) been described from Chile, but may possibly be found there in the future, like *B. longicauda* Daday, 1901, *B. sylvestri* (Daday, 1901), and *B. vallentini* (T. Scott, 1914) (cf. Menu-Marque et al., 2000). Other important groups are the fairy shrimps of the genus *Branchinecta* (cf. Rogers et al., 2008), that are rare in zooplankton assemblages, i.e., mainly restricted to shallow ephemeral ponds (Soto, 1990), and possibly also to oligotrophic and low conductivity water bodies (De los Ríos, 2005, 2008; De los Ríos et al., 2008a, b, c). Also, the presence of Notostraca has been reported from shallow ephemeral pools, although those descriptions do not specify any details (Soto, 1990). The high species richness in shallow ponds, mainly ephemeral water bodies (De los Ríos, 2008; De los Ríos et al., 2008a, b) agrees with the first descriptions of Soto (1990), and that observation agrees with descriptions of Spencer et al. (1999), who described the existence of a direct relationship between surface area and species richness, which is more explicit in temporal pools. The importance of these water bodies and their high species richness is, that this provides more individuals that may disperse into the surrounding water bodies, i.e., that may engage in processes of colonization (Blaustein & Schwartz, 2001). The advantage obviously is, that the ephemeral pools are close together and join in an ecological system that provides facilities for species dispersal into habitats that are environmentally quite similar: a whole assemblage of such shallow water bodies extending over the vast plains (Spencer et al., 2002). This would help explain the marked dispersion of various essentially Patagonian species, such as *Boeckella poppei* (cf. Menu-Marque et al., 2000; Pugh et al., 2002) and *Parabroteas sarsi*, that are widespread in central and southern Patagonia as well as on the sub-Antarctic islands (De los Ríos & Rivera, 2008).

Another important finding is the presence of *Artemia persimilis*, that can coexists with copepods such as harpacticoids (De los Ríos, 2005) or *Boeckella poopoensis* (cf. P. De los Ríos & G. Gajardo, unpubl. data). This situation is rare, because both species do not usually coexist in South American saline lakes (Hurlbert et al., 1984, 1986; Williams et al., 1995; Zúñiga et al., 1999; De los Ríos & Crespo, 2004; De los Ríos, 2005). The species assemblages here found with both calanoids and brine shrimps, are rather similar to those observed on the Altiplano (Campos et al., 1996; De los Ríos, 2005; De los Ríos & Contreras, 2005). Yet, in spite of the salinity gradient, salinity as an exclusive factor has no significant effect as a regulator of zooplankton assemblages (De los Ríos, 2005; Soto & De los Ríos, 2006), though the literature mentions that salinity and trophic status would allegedly be the main regulating factors for zooplankton assemblages in shallow water bodies (Soto & De los Ríos, 2006). This may be caused by combined effects, in which case the scenarios would be: (1) low conductivity and mesotrophic status, supporting a high species richness; (2) a wide conductivity gradient and oligotrophy, with low species richness; and (3) a moderate conductivity with mesotrophic status, also with an accordingly low species richness (Soto & De los Ríos, 2006).

In the sense of community ecology, the existence of a direct relationship between species richness and habitat surface was not reported (Dodson, 1991, 1992; Dodson & Silva-Briano, 2005), nor have relations been found here as those observed for water bodies that were altered through human activities (Dodson et al., 2000, 2005, 2007; Hoffman & Dodson, 2005). If we consider the information available on shallow water bodies, first of all neither surface area nor depth would probably be relevant, due their small size. However,

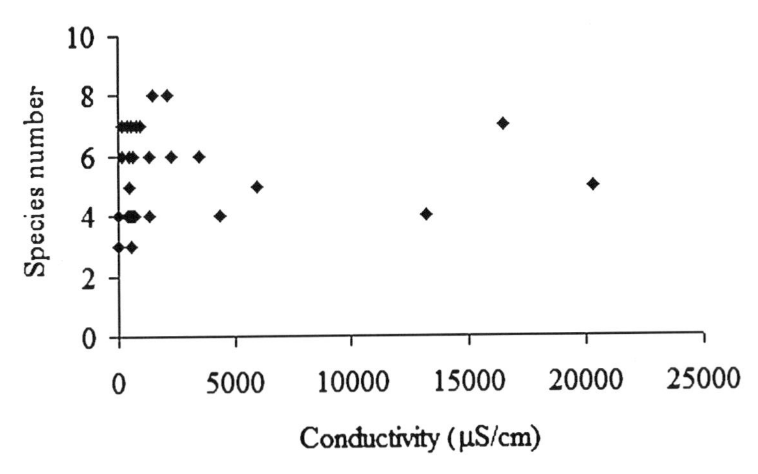

Fig. 38. Graph of the correlation of conductivity versus species richness in zooplankton associations in shallow lakes in southern Patagonia.

it does seem probable the main regulating factors would be conductivity and trophic status, together governing the composition and abundance of the locally present species assemblages (De los Ríos et al., 2008a, b). This situation thus would be different in comparison with Northern Hemisphere lakes, where surface area and depth have a major influence (Dodson et al., 2000, 2005, 2007). In fact, Campos et al. (1996) reported that zooplankton assemblages of southern Patagonian shallow ponds would be similar to those found on the South American Altiplano. The role of conductivity as an isolated factor thus seems to be weak (De los Ríos, 2005; fig. 38), which result would agree with the descriptions of De los Ríos et al. (2008a, b) that indicate the existence of a combined effect of conductivity and trophic status as the primary regulators of zooplankton assemblages.

This kind of zooplankton communities is similar to those of the central Argentinean plains, which harbour numerous shallow ponds (Echaniz et al., 2006; Vignatti et al., 2006). Those ecosystems would have similarities with the southern Chilean Patagonian shallow ponds, due the role of trophic status and conductivity as regulators of zooplankton assemblages. Another important factor is the presence of migratory aquatic birds such as swans, flamingoes, and ducks that use the shallow water bodies as nesting and feeding areas (Soto, 1990; Campos et al., 1996; Araya & Millie, 2001). One of such bird species is the Chilean flamingo, which is an active predator on crustacean zooplankton (López, 1990), probably also in shallow Patagonian water bodies (Soto, 1990). The migratory routes of those birds may well have a role in the dispersal of the plankton species in those areas, as some of those species are widespread in Patagonian water bodies (Araya & Millie, 2005). Thus, aquatic birds would be important both as predators and as a dispersal agent. An an example, widespread species such as *Boeckella gracilipes*, *B. meteoris*, and *B. poopoensis* are reported from sites that are in the migratory routes of the Chilean flamingo (Araya & Millie, 2001).

Also, these water bodies harbour populations of *Parabroteas sarsi*, a large-sized calanoid that is an active predator on small-sized zooplankton (Balseiro & Vega, 1994; Vega, 1996, 1997, 1998). This species is distributed in Chile from 45 to 53°S, while in Argentina it is reported south of 38°S (De los Ríos & Rivera, 2008). *P. sarsi* is relatively ubiquitous, and can be found over a wide gradient conditions, set by the environmental parameters of conductivity and trophic status (Reissig et al., 2004; De los Ríos, 2005; De los Ríos & Contreras, 2005; Soto & De los Ríos, 2006). The scarce ecological studies available are based on observations from northern Argentinean Patagonia (Balseiro & Vega, 1994; Vega, 1997, 1998, 1999; Modenutti et al., 1998) and sub-Antarctic islands (Hansson et al., 1996; Hansson & Tranvik, 1997, 2003; Pugh et al., 2002; Dartnall et al., 2005). In general, these studies mention that juveniles of this species graze on phytoplankton in the first stages of their life cycle, and in later stages change gradually to a diet of small-sized zooplankton: first rotifers and nauplii, and next in the (sub-)adult life stages their preys comprise copepodids and juveniles of cladocerans (fig. 39).

For sub-Antarctic islands, a wide range of environmental conditions has been described (Hansson et al., 1996), that would explain the considerable species richness in crustacean zooplankton reported for their inland waters (Pugh et al., 2002; Dartnall, 2005). The same regulator mechanism reported

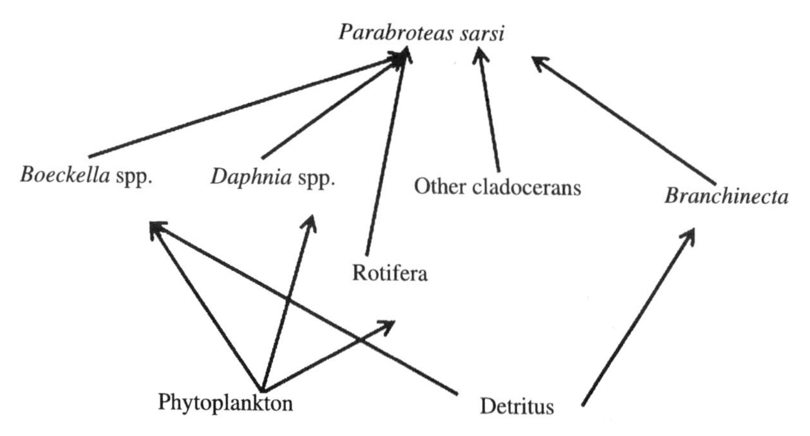

Fig. 39. Trophic interactions involving planktonic crustaceans in shallow Patagonian lagoons.

TABLE X

Conductivity values (in μS/cm) measured for the habitats of crustacean zooplankton species reported from southern Patagonia

Species	Average	Minimum	Maximum
Boeckella gracilipes Daday	3070.1	154.0	20301.0
B. meteoris Kiefer	11887.7	5901.0	16561.0
B. michaelseni (Mrázek)	3761.4	151.0	20301.0
B. poopoensis Marsh	16560.0	No data	No data
B. poppei (Mrázek)	2835.7	151.0	20301.0
Parabroteas sarsi (Ekman)	2972.1	154.0	16561.0
Cyclopoida indet.	2637.3	151.0	16561.0
Chydorus sphaericus (O. F. Müller)	2015.9	151.0	6140.0
Daphnia dadayana Paggi	1295.8	151.0	6140.0
D. obtusa Kurz	1660.2	553.0	3449.0
D. pulex Leydig	2852.8	154.0	20301.0
Neobosmina chilensis (Daday) [= *Eubosmina hagmanni* (Stingelin)]	4499.5	151.0	20301.0
Branchinecta spp.	607.4	420.0	962.0

for sub-Antarctic inland waters (i.e., conductivity and trophic status combined) would affect the zooplankton community structure in shallow lagoons in the continental zone of the adjoining regions (De los Ríos, 2008; De los Ríos et al., 2008a, b). In fact, the ecological reports of Hansson et al. (1996), observed the absence of *Branchinecta* in water bodies of the island South Georgia, and the environmental conditions of these habitats are similar to those in continental water bodies without *Branchinecta* (cf. De los Ríos et al., 2008a, b). More details about the trophic interactions between "entomostracan" species in these habitats would be interesting, mainly involving the potential predation of *Parabroteas sarsi* on other crustacean preys, i.e., juvenile stages of *Branchinecta* and *Artemia* nauplii.

Dominant species

Many of these water bodies show a dominance of calanoids at moderate to low conductivity (Soto & De los Ríos, 2006; table X), the most representative species being *Boeckella gracilipes*, *B. michaelseni*, and *B. poppei* (cf. Menu-Marque et al., 2000; Soto & De los Ríos, 2006). At a mesotrophic level the representative species are *Daphnia ambigua* and *D. dadayana* (cf. Soto & De los Ríos, 2006; De los Ríos, 2008). Finally, at high conductivity the anostracan *Artemia persimilis* can be dominant, even as the exclusive crustacean zooplankton component (Campos et al., 1996), or else coexisting with halophilic copepods (De los Ríos, 2005; P. De los Ríos & G. Gajardo, in press). This situation underscores the role of conductivity as the main determinant of the species structure of the zooplankton assemblages, similar to the situation found in northern Chilean saline waters (De los Ríos, 2005; De los Ríos & Contreras, 2005). In fact, at low to moderate conductivity we find an increase in the fraction of calanoids, i.e., by number of species, in southern Chilean shallow saline lakes and pools (Soto & De los Ríos, 2006). Although the calanoids are dominant, however, the conductivity of these water bodies does not allow the exclusive presence of this group in southern Chilean shallow water bodies (Soto & De los Ríos, 2006), as is observed in northern Chilean saline lakes (De los Ríos & Crespo, 2004). The most representative calanoid species in southern shallow ponds are *Boeckella meteoris*, *B. poopoensis*, *B. poppei*, and *Parabroteas sarsi* (cf. De los Ríos & Contreras, 2005). While we find a trend in calanoid dominance involving the group as a whole, similar trends of dominance can be observed within the group at an increase in conductivity (Soto & De los Ríos, 2006). These observations are similar to existing descriptions of sub-Antarctic water bodies, and thus a similar set of regulating factors is apparently involved, i.e., the combination of conductivity and trophic status (Hansson et al., 1996; Hansson & Tranvik, 1997, 2003; Pugh et al., 2002; Dartnall et al., 2005). A different scenario, however, is observed in central Argentinean shallow water bodies, where calanoids are not necessarily associated with a conductivity gradient (Modenutti et al., 1998), and the main representative species are *Boeckella antiqua* Menu-Marque & Balseiro, 2000, *B. gibbosa*, and *B. gracilis* (cf. Menu-Marque & Balseiro, 2000; Menu-Marque et al., 2000; Trochine et al., 2003). Similar results were reported for saline and subsaline water bodies in New Zealand, where calanoids are also dominant (Hall & Burns, 2001a, b). This wide environmental variability as well as its effects on zooplankton assemblages have also been described for the Plateau of British Columbia in Canada, where it proved possible to explain differences in the abundances of various zooplankton species, including complete absence, according to the ambient salinity and to the concentration of cations, mainly calcium and magnesium (Bos et al., 1996). In that study, however, the various assemblages were characterized by the presence of abundant daphniid populations at low salinity, and *Artemia* at high salinity (Bos et al., 1996). The pattern of abundance of daphniids at moderate to high salinity has been adequately described already for Northern Hemisphere lakes (Frey, 1993).

The web of trophic interactions implies that calanoids, mainly *Boeckella poppei*, play a key role (Bayly & Burton, 1993), because they constitute the main prey for the scarce predators, such as aquatic insects (Arnold & Convey, 1998), and simultaneously are the main grazer/predator on phytoplankton, bacteria, and picoplankton (Almada et al., 2004; Büttler et al., 2005; Allende & Pizarro, 2006). The role of the large-sized calanoid predator, *Parabroteas sarsi*, which predates on rotifers and the juvenile stages of copepods and cladocerans, is important as well (Balseiro & Vega, 1994; Vega, 1996, 1997, 1998, 1999; Dieguez & Balseiro, 1998; Brandl, 2005). Another interesting topic is the role of fairy shrimps, which are detritivorous, consuming decomposing vegetal material, and grazing on the periphyton (Paggi, 1994, 1996b; Pociecha & Dumont, 2008). Unfortunately, there is no more detailed information about trophic interactions in those assemblages, with data mainly lacking on the issues of competition and/or of niche-sharing or niche-segregation between or among the various species in the zooplankton.

Effects of exposure to natural ultraviolet radiation

The water bodies in central and southern Patagonia are exposed to natural ultraviolet radiation, which is enhanced due to the recent depletion of the ozone in the atmosphere at extreme southern latitudes as a result of human activities, notably air pollution with CFC's [chlorofluorocarbon compounds] (cf. Villafañe et al., 2001, 2004; see also previous chapter). As their Argentinean counterparts (Morris et al., 1995), also in Chile these water bodies show high concentrations of dissolved organic carbon (DOC) (De los Ríos, 2003). This condition provides a natural protection against UV radiation, by virtue of the screen effect evoked by DOC (Morris et al., 1995; Marinone et al., 2006), which prevents deep penetration into the water column. Yet, an opposite effect is also possible, because sometimes the interaction between dissolved organic carbon and natural ultraviolet radiation can generate harmful substances such as peroxides and other reactive oxygen compounds (Reche et al., 1998). In this scenario, the zooplankters may have evolved protective strategies, such as the synthesis of photoprotective substances (examples: melanine or carotenoids), or antioxidant compounds (mycosporine-like aminoacids, or ascorbic acid) that can provide protection against direct natural ultraviolet radiation and/or protect against oxidative stress (García et al., 2008). This situation was reported for southern Patagonian water bodies, where the species are mainly calanoids that contain significant quantities of mycosporine-like aminoacids (Villafañe et al., 2001; Tartarotti et al., 2004; García et al., 2008), or else the presence of melanine in *Daphnia dadayana* (cf. Paggi, 1999; De los Ríos,

2005). For shallow Patagonian water bodies a different situation has been described, because these waters are exposed to strong winds (Soto et al., 1994) that generate a strong mixing of the entire water column (Zagarese et al., 1998; Soto, 2002). In this scenario, the species of zooplankton have a different kind of tolerance to ultraviolet radiation, and the calanoids are more tolerant to exposure to natural ultraviolet radiation, mainly under an oligotrophic regime (De los Ríos, 2005). These results are similar to observations reported for Antarctic and sub-Antarctic environments that share similar species, such as *Boeckella poppei* (cf. Rocco et al., 2002).

The exposure to natural ultraviolet radiation and its effects have been described for Arctic shallow ponds, where, as a function of geophysical characteristics, mainly altitude above sea level, as well as chemical parameters, specifically DOC concentration, a different gradient of the penetration of natural ultraviolet radiation was reported (Rautio, 2002). Here, the representative species are melanized species of *Daphnia* (cf. Rautio & Korkhola, 2002a, b). A similar situation would apply to northern Patagonia, where the high concentration of dissolved organic carbon would provide protection against exposure to natural ultraviolet radiation and, at mesotrophic levels, melanized *D. dadayana* can be dominant (Soto & De los Ríos, 2006). Experimental evidence demonstrated that this last species is not tolerant to ultraviolet radiation exposure under an oligotrophic regime, in contrast to calanoids (De los Ríos, 2005), and this condition would agree with the results of De los Ríos (2003) who indicated that cladocerans are abundant in mesotrophic lakes with high dissolved organic carbon content, that builds a natural screen against ultraviolet radiation.

ACKNOWLEDGEMENTS

The author expresses his gratitude to Dr. J. C. von Vaupel Klein, who gave encouragement and support for realizing this work; he is also grateful to Prof. Santiago Peredo for his valuable comments and suggestions, as well as to B.Sc. Reinaldo Rivera, B.Sc. Marcela Galindo, B.Sc. Marilyn González, and B.Sc. Esteban Quinan, who, as students and graduates, provided important assistance for the preparation of the present book.

Also, the valuable assistance of Brigitte Lechner is recognized, who supplied me with the first reports on Chilean crustacean zooplankton.

REFERENCES

ACEITUNO, P., 1997. Aspectos generales del clima en el Altiplano Sudamericano. In: Actas del II. Simposio Internacional de Estudios Altiplánicos, Arica, Chile: 63-70.

ADAMOWICZ, S., S. MENU-MARQUE, P. HEBERT & A. PURVIS, 2007. Molecular systematics and patterns of morphological evolution in the Centropagidae (Copepoda: Calanoida) of Argentina. Biol. Journ. Linn. Soc. London, **90**: 279-292.

ALLENDE, L. & H. PIZARRO, 2006. Top-down control on plankton components in an Antarctic pond: experimental approach to the study of low-complexity food webs. Polar Biol., **29**: 893-901.

ALMADA, P., L. ALLENDE, G. TELL & I. IZAGUIRRE, 2004. Experimental evidence of the grazing impact of *Boeckella poppei* on phytoplankton in a maritime Antarctic lake. Polar Biol., **28**: 39-46.

ALONSO, C., V. RICCI, J. P. BARRIGA, M. A. BATTINI & H. ZAGARESE, 2004. Surface avoidance by freshwater zooplankton: field evidence on the role of ultraviolet radiation. Limnol. Oceanogr., **49**: 225-232.

ARAYA, B. & G. MILLIE, 2005. Guía de campo de las aves de Chile (5th ed.): 1-406. (Editorial Universitaria, Santiago de Chile).

ARAYA, J. M. & L. R. ZÚÑIGA, 1985. Manual taxonómico del zooplancton lacustre de Chile. Boln. Limnológico, Universidad Austral de Chile, **8**: 1-110.

ARNOLD, R. J. & P. CONVEY, 1998. The life history of the diving beetle, *Lancetes angusticollis* (Curtis) (Coleoptera: Dytiscidae), on sub-Antarctic South Georgia. Polar Biol., **20**: 153-160.

BALSEIRO, E. G., B. E. MODENUTTI & C. P. QUEIMALIÑOS, 2001. Feeding of *Boeckella gracilipes* (Copepoda, Calanoida) on ciliates and phytoflagellates in an ultraoligotrophic Andean lake. Journ. Plankt. Res., **23**: 849-857.

BALSEIRO, E. G. & M. VEGA, 1994. Vulnerability of *Daphnia middendorffiana* to *Parabroteas sarsi* predation: the role of the tail spine. Journ. Plankt. Res., **16**: 783-793.

BAYLY, I. A. E., 1972. Salinity tolerance and osmotic behaviour of animals in athalassic saline waters and marine hypersaline waters. Annu. Rev. Ecol. Syst., **3**: 233-268.

— —, 1992a. The non-marine Centropagidae (Copepoda, Calanoida) of the world. Guides to the Identification of the Microinvertebrates of the Continental Waters of the World, **2**: 1-30. (SPB Academic Publishers, Amsterdam).

— —, 1992b. Fusion of the genera *Boeckella* and *Pseudoboeckella* and a revision of their species from South America and Subantarctic islands. Rev. Chilena Hist. nat., **65**: 17-63.

— —, 1993. The fauna of athalassic saline waters in Australia and the Altiplano of South America: comparison and historical perspectives. Hydrobiologia, **267**: 225-231.

— —, 1995. Distinctive aspects of the zooplankton of large lakes in Australasia, Antarctica and South America. Mar. freshwat. Res., **46**: 1109-1120.

BAYLY, I. A. E. & R. H. BURTON, 1993. Beaver Lake, Greater Antarctica, and its population of *Boeckella poppei* (Mrázek) (Copepoda: Calanoida). Verh. intern. Verein. angew. Limnol., **25**: 975-978.

BLAUSTEIN, L. & S. S. SCHWARTZ, 2001. Why study ecology of temporary ponds? Israel Journ. Ecol. Evol., **47**: 303-312.

BOS, D. G., B. F. CUMMING, C. E. WATTERS & J. P. SMOL, 1996. The relationship between zooplankton, conductivity and lake-water ionic composition in 111 lakes from the Interior Plateau of British Columbia, Canada. Intern. Journ. Salt Lake Res., **5**: 1-15.

BRANDL, Z., 2005. Freshwater copepods and rotifers: predators and their prey. Hydrobiologia, **546**: 475-489.

BREHM, V., 1935a. Mitteilungen von den Forschungsreisen Prof. Rahms. Mitteilung I. Zwei neue Entomostraken aus der Wüste Aracama. Zool. Anz., **111**: 279-284.

— —, 1935b. Mitteilungen von den Forschungsreisen Prof. Rahms. Mitteilung II. Gibt es in der chilenischen Region Diaptomiden? *Diaptomus diabolicus* nov. spec. Zool. Anz., **112**: 9-13.

— —, 1935c. Mitteilungen von den Forschungsreisen Prof. Rahms. Mitteilung III. Copepoden aus Cajon del Plomo in der Kordillere von Santiago, 3.330 m. Zool. Anz., **112**: 73-79.

— —, 1935d. Mitteilungen von den Forschungsreisen Prof. Rahms. Mitteilung IV. Über eine mit *Pseudoboeckella valentini* Scott nächstverwandte *Pseudoboeckella* aus Chile: *Pseudoboeckella gibbosa* sowie über eine weitere neue *Pseudoboeckella* und *Alona*. Zool. Anz., **112**: 116-123.

— —, 1936. Mitteilungen von den Forschungsreisen Prof. Rahms. Mitteilung VI. Über die Cladocerenfauna des Titicaca und über einige neue Fundstellen bereits bekannter Copepoden. Zool. Anz., **114**: 157-159.

— —, 1937. Eine neue *Boeckella* aus Chile. Zool. Anz., **118**: 304-307.

BRENDONCK, L., 1996. Diapause, quiescence, hatching requirements: what we learn from large freshwater branchiopods (Crustacea: Branchiopoda: Anostraca, Notostraca, Conchostraca). Hydrobiologia, **320**: 85-97.

BRTÉK, D. & G. MURA, 2000. Revised key to families and genera of the Anostraca with notes on their geographical distribution. Crustaceana, **73** (9): 1037-1088.

BUSTAMANTE, R. O., A. A. GREZ & J. A. SIMONETTI, 2006. Efectos de la fragmentación del bosque maulino sobre la abundancia y diversidad de especies nativas. In: A. A. GREZ, J. A. SIMONETTI & R. O. BUSTAMENTE (eds.), Biodiversidad y patrones en ambientes fragmentados de Chile: patrones y procesos a diferentes escalas: 83-98. (Editorial Universitaria, Santiago de Chile).

BUTLER, H., A. ATKINSON & M. GORDON, 2005. Omnivory and predation impact of the calanoid copepod *Boeckella poppei* in a maritime Antarctic lake. Polar Biol., **28**: 815-822.

CABRERA, S., M. LÓPEZ & B. TARTAROTTI, 1997. Phytoplankton response to ultraviolet radiation in a high altitude Andean lake: short- versus long-term effects. Journ. Plankt. Res., **19**: 1565-1582.

CABRERA, S. & V. MONTECINO, 1987. Productividad primaria en sistemas límnicos. Archivos Biol. Med. exp., **20**: 105-116.

CAMPOS, H., 1984. Limnological study of Araucanian lakes (Chile). Verh. intern. Verein. angew. Limnol., **22**: 1319-1327.

CAMPOS, H., J. ARENAS, W. STEFFEN, C. ROMÁN & G. AGÜERO, 1982. Limnological study of Lake Ranco (Chile): morphometry, physics and plankton. Arch. Hydrobiol., **94**: 137-171.

CAMPOS, H., E. BUCAREY & J. ARENAS, 1974. Estudios limnológicos del lago Riñihue y río Valdivia (Chile). Boln. Soc. Biol. Concepción, **48**: 47-67.

CAMPOS, H., D. SOTO, O. PARRA, W. STEFFEN & G. AGÜERO, 1996. Limnological studies of Amarga lagoon, Chile: a saline lake in Patagonia, South America. Intern. Journ. Salt Lake Res., **4**: 301-314.

CAMPOS, H., D. SOTO, W. STEFFEN, G. AGÜERO, O. PARRA & L. ZÚÑIGA, 1994a. Limnological studies in Lake del Toro, Chilean Patagonia. Arch. Hydrobiol., **99**: 199-215.

— —, — —, — —, — —, — — & — —, 1994b. Limnological studies in lake Sarmiento, a subsaline lake from Chilean Patagonia. Arch. Hydrobiol., **99**: 217-234.

CAMPOS, H., W. STEFFEN, G. AGÜERO, O. PARRA & L. ZÚÑIGA, 1983. Limnological studies in Lake Villarrica. Morphometry, physics, chemistry and primary productivity. Arch. Hydrobiol., (Suppl.) **71**: 37-67.

— —, — —, — —, — — & — —, 1987a. Limnology of Lake Riñihue. Limnológica, **18**: 339-357.

— —, — —, — —, — — & — —, 1987b. Estudios limnológicos en el lago Caburgua (Chile). Gayana, Botánica, **44**: 61-84.

— —, — —, — —, — — & — —, 1988. Limnological study of Lake Llanquihue (Chile): morphometry, physics, chemistry and primary productivity. Arch. Hydrobiol., (Suppl.) **81**: 37-67.

— —, — —, — —, — — & — —, 1989. Estudios limnológicos en el lago Puyehue (Chile): morfometría, factores físicos y químicos, plancton y productividad primaria. Medio Ambiente, **10**: 36-53.

— —, — —, — —, — — & — —, 1990. Limnological study of Lake Todos los Santos (Chile): morphometry, physics, chemistry and primary productivity. Arch. Hydrobiol., (Suppl.) **117**: 453-484.

— —, — —, — —, — — & — —, 1992a. Limnological study of Lake Ranco (Chile). Limnológica, **22**: 337-353.

— —, — —, — —, — — & — —, 1992b. Limnological studies of Lake Rupanco (Chile): morphometry, physics, chemistry and primary productivity. Arch. Hydrobiol., (Suppl.) **90**: 85-113.

CAMPOS, V., 1997. Microorganismos de ambientes extremos: Salar de Atacama, Chile. In: Actas del II. Simposio Internacional de Estudios Altiplánicos, Arica, Chile: 143-147.

CHANG, K. H. & T. HANAZATO, 2003a. Vulnerability of cladoceran species to predation by the copepod *Mesocyclops leuckarti*: laboratory observations on the behavioural interactions between predator and prey. Freshwat. Biol., **48**: 476-484.

— — & — —, 2003b. Seasonal and reciprocal succession and cyclomorphosis of two *Bosmina* species (Cladocera, Crustacea) co-existing in a lake: their relationship with invertebrate predators. Journ. Plankt. Res., **25**: 141-150.

— — & — —, 2005a. Impact of selective predation by *Mesocyclops pehpeiensis* on a zooplankton community: experimental analysis using mesocosm. Ecol. Res., **20**: 726-723.

— — & — —, 2005b. Prey handling time and ingestion probability for *Mesocyclops* sp. Predation on small cladoceran species *Bosmina longirostris*, *Bosminopsis deitersi* and *Scapholeberis mucronata*. Limnology, **6**: 29-34.

CHANG, K. H., T. NAGATA & T. HANAZATO, 2004. Direct and indirect impacts of predation by fish on the zooplankton community: an experiment using tanks. Limnology, **5**: 121-124.

CHONG, G., 1988. The Cenozoic saline deposit of the Chilean Andes between 18°00 and 27°00 south latitude. Lecture Notes on Earth Sciences, **17**: 137-151.

CRESPO, J. & P. DE LOS RÍOS, 2004. A new locality for *Artemia* Leach, 1819 (Branchiopoda, Anostraca) in Chile. Crustaceana, **77** (2): 245-247.

DADAY, E., 1902. Beiträge zur Kenntniss der Süsswasser-Mikrofauna von Chile. Természetr. Fütezek, **25**: 436-447.

DANA, G. L., C. J. FOLEY, G. L. STARRET, W. M. PERRY & J. M. MELACK, 1988. In situ hatching of *Artemia monica* cysts in hypersaline Mono Lake, California. Hydrobiologia, **158**: 183-190.

DANA, G. L., R. JELLISON & J. M. MELACK, 1990. *Artemia monica* cyst production and recruitment in Mono Lake, California, USA. Hydrobiologia, **187**: 233-243.

DANA, G. L., R. JELLISON, J. M. MELACK & G. L. STARRET, 1993. Relationships between *Artemia monica* life history characteristics and salinity. Hydrobiologia, **263**: 129-143.

DANA, G. L. & P. LENZ, 1986. Effects of increasing salinity on an *Artemia* population from Mono Lake, California. Oecologia, **68**: 428-436.

DARTNALL, J. G., 2005. Freshwater invertebrates of subantarctic South Georgia. Journ. nat. Hist., London, **39**: 3321-3342.

DEEVEY, E. S. & G. B. DEEVEY, 1971. The American species of *Eubosmina* Seligo (Crustacea, Cladocera). Limnol. Oceanogr., **16**: 201-218.

DE LOS RÍOS, P., 2003. Efectos de las disponibilidades de recursos energéticos, estructurales y de protección sobre la distribución y abundancia de crustáceos zooplanktónicos lacustres chilenos: 1-163. (Doctoral Thesis, Austral University of Chile, Science Faculty, Valdivia).

— —, 2005a. Richness and distribution of zooplanktonic crustacean species in Chilean altiplanic and southern Patagonia ponds. Polish Journ. envir. Stud., **14**: 817-822.

— —, 2005b. Survival of pigmented freshwater zooplankton exposed to artificial ultraviolet radiation and two levels of dissolved organic carbon. Polish Journ. Ecol., **53**: 113-116.

— —, 2008. A null model for explain crustacean zooplankton species associations in central and southern Patagonian inland waters. Anales Inst. Patagonia, **36**: 25-33.

DE LOS RÍOS, P., P. ACEVEDO, R. RIVERA & G. ROA, 2008. Comunidades de crustáceos litorales de humedales del norte de la Patagonia chilena (38°S): rol potencial de la exposición a la radiación ultravioleta. In: A. V. VOLPEDO & L. FERNÁNDEZ (eds.), Efecto de los cambios globales sobre la biodiversidad: 209-218. (CYTED Programa Iberoamericano de Ciencia y Tecnología para el Desarrollo, Red 406RT0285).

DE LOS RÍOS, P. & P. CONTRERAS, 2005. Salinity level and occurrence of centropagid copepods (Crustacea, Copepoda, Calanoida) in shallow lakes in Andes mountains and Patagonian plains, Chile. Polish Journ. Ecol., **53**: 445-450.

DE LOS RÍOS, P. & J. CRESPO, 2004. Salinity effects on the abundance of *Boeckella poopoensis* (Copepoda, Calanoida) in saline ponds in the Atacama desert, northern Chile. Crustaceana, **77** (4): 417-423.

DE LOS RÍOS, P. & G. GAJARDO, 2004. Letter to editor: biological bases for a potential management of Chilean *Artemia* (Crustacea, Anostraca) populations. Rev. Chilena Hist. nat., **77**: 3-4.

— — & — —, in press. Potential heterogeneity in crustacean zooplankton assemblages in southern Chilean saline lakes. Brazilian Journal of Biology.

DE LOS RÍOS, P., E. HAUENSTEIN, P. ACEVEDO & X. JAQUE, 2007. Littoral crustaceans in mountain lakes of Huerquehue National Park (38°S, Araucania Region, Chile). Crustaceana, **80** (4): 401-410.

DE LOS RÍOS, P. & N. RIVERA, 2007. *Branchinecta* (Branchiopoda, Anostraca) as bioindicator of oligotrophic and low conductivity shallow water bodies in southern Chilean Patagonia. Anales Inst. Patagonia, **35**: 15-20.

— — & — —, 2008. On the geographic distribution of *Parabroteas sarsi* (Mrázek, 1901) (Copepoda, Calanoida). Anales Inst. Patagonia, **36**: 75-78.

DE LOS RÍOS, P., N. RIVERA & M. GALINDO, 2008a. The use of null models to explain crustacean zooplankton associations in shallow water bodies of the Magellan region, Chile. Crustaceana, **81** (10): 1219-1228.

DE LOS RÍOS, P. & G. ROA, 2010. Crustacean species assemblages in mountain shallow ponds: Parque Cañi (38°S, Chile). Zoologia, Curitiba, **27**: 81-86.

DE LOS RÍOS, P., D. C. ROGERS & N. RIVERA, 2008b. *Branchinecta gaini* Daday, 1910 (Branchiopoda, Anostraca) as a bioindicator of oligotrophic and low conductivity shallow ponds in southern Chilean Patagonia. Crustaceana, **81** (9): 1025-1034.

DE LOS RÍOS, P. & M. ROMERO-MIERES, 2009. Littoral crustaceans in lakes of Conguillío National Park (38°S), Araucania region, Chile. Crustaceana, **82** (1): 117-119.

DE LOS RÍOS, P., M. ROMERO-MIERES & E. HAUENSTEIN, in press. Zooplankton assemblages in lakes and ponds of Arauco and Malleco Provinces (37-38°S, Chile). Brazilian Journal of Biology.

DE LOS RÍOS, P. & D. SOTO, 2005. Survival of two species of crustacean zooplankton under two chlorophyll concentrations and protection or exposure to natural ultraviolet radiation. Crustaceana, **78** (2): 163-169.

— — & — —, 2006. Effects of the availability of energetic and protective resources on the abundance of daphniids (Cladocera, Daphniidae) in Chilean Patagonian lakes (39°-51°S). Crustaceana, **79** (1): 23-32.

— — & — —, 2007a. Temporal and spatial variations in trophic status and daphniids (Crustacea) dominance in Llanquihue lake (41°S Chile). Polish Journ. Ecol., **55**: 191-193.

— — & — —, 2007b. Crustacean (Copepoda and Cladocera) zooplankton richness in Chilean Patagonian lakes. Crustaceana, **80** (3): 285-296.

DE LOS RÍOS, P. & L. R. ZÚÑIGA, 2000. Comparación biométrica del lóbulo frontal en poblaciones americanas de *Artemia* (Anostraca, Artemiidae). Rev. Chilena Hist. nat., **73**: 31-38.

DEMERGASSO, C., G. CHONG, P. GALLEGUILLOS, L. ESCUDERO, M. MARTÍNEZ-ALONSO & I. ESTEVE, 2003. Tapetes microbianos del Salar de Llamará, norte de Chile. Rev. Chilena Hist. nat., **76**: 485-499.

DIAZ, S., C. CAMILLÓN, G. DEFERRARI, H. FUENZALIDA, R. ARMSTRONG, C. BOOTH, A. PALADANI, S. CABRERA, C. CASICCIA, C. LOVENGREEN, J. PEDRONI, A. ROSALES, H. ZAGARESE & M. VERNET, 2006. Ozone and UV radiation over southern South America: climatology and anomalies. Photochem. Photobiol., **82**: 834-843.

DIÉGUEZ, M. C. & E. G. BALSEIRO, 1998. Colony size in *Conochilus hippocrepis*: defensive adaptation to predator size. Hydrobiologia, **387**: 421-425.

DODSON, S. I., 1991. Species richness of crustacean zooplankton in European lakes of different sizes. Verh. intern. Verein. theor. angew. Limnol., **24**: 1223-1229.

— —, 1992. Predicting crustacean zooplankton species richness. Limnol. Oceanogr., **37**: 848-856.

DODSON, S. I., S. E. ARNOTT & K. L. COTTINGHAM, 2000. The relationship in lake communities between primary productivity and species richness. Ecology, **81**: 2662-2679.

DODSON, S. I., W. R. EVERHAST, A. K. JANDL & S. J. KRAUSKOPF, 2007. Effect of watershed land use and lake age on zooplankton species richness. Hydrobiologia, **579**: 393-399.

DODSON, S. I., R. A. LILLIE & S. WILLE-WOLF, 2005. Land use, water chemistry, aquatic vegetation and zooplankton community structure of shallow lakes. Ecol. applic., **15**: 1191-1198.

DODSON, S. I. & M. SILVA-BRIANO, 2005. Crustacean zooplankton species richness and associations in reservoirs and ponds of Aguas Calientes, Mexico. Hydrobiologia, **325**: 163-172.

DOMÍNGUEZ, P., 1971. Nota sobre *Moina micrura* Kurz. Anales Mus. Hist. nat. Valparaíso, **4**: 353-358.

— —, 1973. Contribución al estudio de los cladóceros chilenos. I. Cladóceros del lago Chungará. Noticiario mens. Mus. Nac. Hist. nat. Santiago de Chile, **17** (201-202): 3-10.

DOMÍNGUEZ, P. & L. ZÚÑIGA, 1976. Análisis fenológico de los cladóceros limnéticos (Crustacea: Entomostraca) de la laguna El Plateado (Valparaíso). Anales Mus. Hist. nat. Valparaíso, **9**: 35-44.

— — & — —, 1979. Perspectiva temporal de la Entomostraca fauna limnética de lago Ranco (Valdivia, Chile). Anales Mus. Hist. nat. Valparaíso, **12**: 53-58.

DYER, B., 2001. Revisión sistemática de los pejerreyes de Chile (Teleostei, Atheriniformes). Estudios Oceanológicos, **19**: 99-127.

ECHANIZ, S. A., A. M. VIGNATTI, S. J. DE PAGGI, J. C. PAGGI & A. PILATI, 2006. Zooplankton seasonal abundance of South American saline shallow lakes. Intern. Revue ges. Hydrobiol., **91**: 86-100.

EKMAN, S., 1900. Cladoceren aus Patagonien gesammelt von der schwedischen Expedition nach Patagonien, 1899. Zool. Jahrb., (Syst.) **14**: 62-84.

ELSER, J. J., D. R. DOBBERFUHL, N. A. MACLAY & J. H. SCHAMPEL, 1996. Organism size, life history, and N : P stoichiometry. Toward a unified view of cellular and ecosystem processes. Bioscience, **46**: 674-684.

ELSER, J. J., W. F. FAGAN, R. F. DENNO, D. R. DOBBERFUHL, A. FOLARIN, A. HUBERTY, S. INTERLANDI, S. S. KILHAM, E. MCCAULEY, K. L. SCHULZ, E. H. SIEMANN & R. W. STERNER, 2000. Nutritional constraint in terrestrial and freshwater food webs. Nature, London, **408**: 578-580.

FREY, D. G., 1993. The penetration of cladocerans into saline waters. Hydrobiologia, **267**: 233-248.

GAJARDO, G. M., N. COLIHUEQUE, M. PARRAGUEZ & P. SORGELOOS, 1998. International study on *Artemia* VIII. Morphologic differentiation and reproductive isolation of *Artemia* populations from South America. Intern. Journ. Salt Lake Res., **7**: 133-151.

GAJARDO, G. M., J. CRESPO, A. TRIANTAFYLLIDIS, A. TZIKA, A. D. BAXEVANIS, I. KAPPAS & T. J. ABATZOPOULOS, 2004. Species identification of Chilean *Artemia* populations based on mitochondrial DNA RFLP analysis. Journ. Biogeogr., **31**: 547-555.

GAJARDO, G. M., M. DA CONCEICAO, L. WEBER & J. A. BEARDMORE, 1995. Genetic variability and interpopulational differentiation of *Artemia* strains of South America. Hydrobiologia, **302**: 21-29.

GAJARDO, G. M., R. WILSON & O. ZÚÑIGA, 1992. Report on the occurrence of *Artemia* in a saline deposit of the Chilean Andes (Branchiopoda, Anostraca). Crustaceana, **63** (2): 169-174.

GARCÍA, P. E., A. P. PÉREZ, M. C. DIÉGUEZ, M. A. FERRARO & H. E. ZAGARESE, 2008. Dual control of the levels of photoprotective compounds by ultraviolet radiation and temperature in the freshwater copepod *Boeckella antiqua*. Journ. Plankt. Res., **30**: 817-827.

GELLER, W., 1992. The temperature stratification and related characteristics of Chilean lakes in midsummer. Aquat. Sci., **54**: 37-57.

GIBBONS, J., Y. A. VILINA & J. CÁRCAMO, 2007. Distribución y abundancia de cisne coscoroba (*Coscoroba coscoroba*), cisne de cuello negro (*Cygnus melanocoryphus*) y del flamenco chileno (*Phoenicopterus chilensis*) en la región de Magallanes. Anales Inst. Patagonia, **35**: 53-68.

GILLOOLY, G. F. & S. I. DODSON, 2000. Latitudinal patterns in the size distribution and seasonal dynamics of New World, freshwater cladocerans. Limnol. Oceanogr., **45**: 22-30.

GLIWICZ, M., 2003. Between hazards of starvation and risk of predation: the ecology of offshore animals: 1-379. (International Ecology Institute, Nordbünte, Oldendorf, Luhe, Germany).

GREEN, A. J., M. I. SÁNCHEZ, F. AMAT, J. FIGUEROLA, F. HONTORIA, O. RUIZ & F. HORTAS, 2005. Dispersal of invasive and native brine shrimps *Artemia* (Anostraca) via waterbirds. Limnol. Oceanogr., **50**: 737-742.

GREZ, A. A. & M. A. BUSTAMANTE-SÁNCHEZ, 2006. Aproximaciones experimentales en estudios de fragmentación. In: A. A. GREZ, J. A. SIMONETTI & R. O. BUSTAMENTE (eds.), Biodiversidad y patrones en ambientes fragmentados de Chile: patrones y procesos a diferentes escalas: 13-16. (Editorial Universitaria, Santiago de Chile).

GROSJEAN, M., M. A. GEYH, B. MESSERLI & U. SCHOTTERER, 1995. Late-glacial and early Holocene lake sediments, ground water formation and climate in the Atacama Altiplano 22-24°S. JOURN. Paleolimnol., **14**: 241-252.

GROSJEAN, M., L. NÜÑEZ, I. CARTAGENA & B. MESSERLI, 1996. Mid-holocene climate and culture change in the Atacama desert, northern Chile. Quaternary Res., New York, **48**: 239-246.

GROSJEAN, M., B. L. VALERO-GARCÉS, M. A. GEYH, B. MESSERLI, U. SCHOTTERER, H. SCHREIER & K. KELTS, 1997. Mid- and late Holocene limnogeology of Laguna del Negro Francisco, northern Chile, and its palaeoclimatic implications. The Holocene, **7**: 151-159.

GUILDFORD, S. J. & R. E. HECKY, 2000. Total nitrogen, total phosphorus and nutrient limitation in lakes and oceans: is there a common relationship? Limnol. Oceanogr., **45**: 1213-1223.

GUTIÉRREZ-AGUIRRE, M. A., E. SUÁREZ-MORALES, A. CERVANTES-MARTÍNEZ, M. ELÍAS-GUTIÉRREZ & D. PREVIATTELLI, 2006. The Neotropical species of *Mesocyclops* (Copepoda, Cyclopoida): an upgraded identification key and comments on selected taxa. Journ. nat. Hist., London, **40**: 549-570.

HALL, C. J. & C. W. BURNS, 2001a. Effects of salinity and temperature on survival and reproduction of *Boeckella hamata* (Copepoda, Calanoida) from a periodically brackish lake. Journ. Plankt. Res., **23**: 97-103.

— — & — —, 2001b. Hatching of *Boeckella hamata* (Copepoda: Calanoida) resting eggs from sediments of a tidally influenced lake. New Zealand Journ. mar. freshwat. Res., **35**: 235-238.

— — & — —, 2002. Mortality and growth responses of *Daphnia carinata* to increases in temperature and salinity. Freshwat. Biol., **47**: 451-468.

HAMMER, U. T. & S. H. HURLBERT, 1992. Is the absence of *Artemia* determined by the presence of predators or by lower salinity in some saline waters. In: R. D. ROBARTS & M. L. BOTHWELL (eds.), Aquatic ecosystems in semi-arid regions: implications for resource management. N.H.R.I. Symposium Series, **7**: 91-102. (Environment Canada, Saskatoon).

HANN, B. J., 1986. Revision of the genus *Daphniopsis* Sars, 1903 (Cladocera-Daphnidae) and a description of *Daphniopsis chilensis*, new species, from South America. Journ. Crust. Biol., **6**: 246-263.

HANSSON, L.-A., H. J. DARTNALL, J. C. ELLIS-EVANS, H. MACALISTER & L. J. TRANVIK, 1996. Variations in physical, chemical and biological components in Subantarctic lakes of South Georgia. Ecography, **19**: 393-404.

HANSSON, L.-A. & L. TRANVIK, 1997. Algal species composition and phosphorus recycling at contrasting grazing pressure: an experimental study in sub-Antarctic lakes with two trophic levels. Freshwat. Biol., **37**: 45-53.

— — & — —, 2003. Food webs in sub-Antarctic lakes: a stable isotope approach. Polar Biol., **26**: 783-788.

HAUENSTEIN, E., M. GONZÁLEZ, F. PEÑA-CORTÉS & A. MUÑOZ-PEDREROS, 2002. Clasificación y caracterización de la flora y vegetación de los humedales de la costa de Toltén (IX región, Chile). Gayana, Botánica, **59**: 87-100.

HENRÍQUEZ, J. M., 2004. Estado de la turba esfagnosa en Magallanes. In: D. E. BLANCO & V. M. DE LA BALSE (eds.), Los turbales de la Patagonia. Wetlands International, Publicación, Buenos Aires, Argentina, **19**: 93 106. [In Spanish with English summary.]

HOFFMAN, M. D. & S. I. DODSON, 2005. Land use, primary productivity and lake area as descriptors of zooplankton diversity. Ecology, **86**: 255-261.

HOFFMEYER, M. S., 2004. Decadal change in zooplankton seasonal succession in the Bahia Blanca estuary, Argentina, following introduction of two zooplankton species. Journ. Plankt. Res., **26**: 181-189.

HURLBERT, S. H., 1982. Limnological studies of flamingo diets and distributions. Nat. geogr. Soc., Res. Rep., **14**: 351-356.

HURLBERT, S. H. & C. C. Y. CHANG, 1963. Ornitholimnology: effects of grazing by the Andean flamingo (*Phoenicoparrus andinus*). Proc. nat. Acad. Sci., U.S.A., **80**: 4766-4769.

HURLBERT, S. H. & J. O. KEITH, 1979. Distribution and spatial patterning of flamingos in the Andean Altiplano. The Auk, **96**: 328-324.

HURLBERT, S. H., W. LOAYZA & T. MORENO, 1986. Fish-flamingo-plankton interactions in the Peruvian Andes. Limnol. Oceanogr., **31**: 457-468.

HURLBERT, S. H., M. LÓPEZ & J. KEITH, 1984. Wilson's phalarope in the central Andes and its interaction with the Chilean flamingo. Rev. Chilena Hist. nat., **57**: 47-57.

JEPPENSEN, E., T. L. LAURIDSEN, S. F. MITCHELL & C. W. BURNS, 1997. Do zooplanktivorous fish structure the zooplankton communities in New Zealand lakes? New Zealand Journ. mar. freshwat. Res., **31**: 163-173.

JEPPENSEN, E., T. L. LAURIDSEN, S. F. MITCHELL, K. CHRISTOFFERSSEN & C. W. BURNS, 2000. Trophic structure in the pelagial of 25 shallow New Zealand lakes: changes along nutrient and fish gradients. Journ. Plankt. Res., **22**: 951-968.

JIMENEZ-MELERO, R., G. PARRA, S. SOUISSI & F. GUERRERO, 2007. Post-embryonic developmental plasticity of *Arctodiaptomus salinus* (Copepoda: Calanoida) at different temperatures. Journ. Plankt. Res., **29**: 553-567.

JIMENEZ-MELERO, R., B. SANTER & F. GUERRERO, 2005. Embryonic and naupliar development of *Eudiaptomus gracilis* and *Eudiaptomus graciloides* at different temperatures: comments on individual variability. Journ. Plankt. Res., **27**: 1175-1187.

KAMJUNKE, N., B. VOGT & S. WÖLFL, 2009. Trophic interactions of the pelagic ciliate *Stentor* spp. in North Patagonian lakes. Limnologica, **39**: 107-114.

KELLER, B. & D. SOTO, 1998. Hydrogeologic influences on the preservation of *Orestias ascotanensis* at Salar de Ascotán, northern Chile. Rev. Chilena Hist. nat., **71**: 147-156.

KIEFER, V., 1936. Mitteilungen von den Forschungsreisen Prof. Rahms. Mitt. VIII. Cyclopiden. Zool. Anz., **115**: 244-249.

KORÍNEK, W. & L. VILLALOBOS, 2003. Two South American endemic species of *Daphnia* from high Andean lakes. Hydrobiologia, **490**: 107-123.

KRŠINIĆ, F., M. CARIĆ, D. VILČIĆ & I. CIGLENEČKI, 2000. The calanoid copepod *Acartia italica* Steuer, phenomenon in the small saline Lake Rogoznica (eastern Adriatic coast). Journ. Plankt. Res., **22**: 1441-1464.

KULL, C. & M. GROSJEAN, 1998. Albedo changes, Milankovitch forcing, and late quaternary climate changes in the central Andes. Clim. Dynam., **14**: 871-881.

LA BARBERA, M. C. & P. KILHAM, 1974. The chemical ecology of copepod distribution in the lakes of East and Central Africa. Limnol. Oceanogr, **19**: 459-465.

LAURION, I., M. VENTURA, J. CATALÁN, R. PSENNER & R. SOMMARUGA, 2000. Attenuation of ultraviolet radiation in mountain lakes: factors controlling the among and within lake variability. Limnol. Oceanogr., **45**: 1274-1268.

LOCASCIO DE MITROVICH, C. & S. MENU-MARQUE, 2001. A new *Diacyclops* (Copepoda, Cyclopoida, Cyclopidae) from northwestern Argentina. Hydrobiologia, **453/454**: 533-538.

LÖFFLER, H., 1962. Zür Systematik und Ökologie der chilenischen Süsswasser-entomostraken. Beitr. Neotr. Fauna, **2**: 145-222. [Printed in paper: 1961.]

LÓPEZ, M., 1990. Alimentación de flamencos altiplánicos con énfasis en *Phoenicoparrus andinus* (Philipii) en el Salar de Cartote, Chile. In: Actas del I. Taller Internacional de Especialistas en Flamencos Sudamericanos: 84-89. (Corporación Nacional Forestal Chile and Zoological Society of New York).

— —, 1997. Comunidades bentónicas de lagunas altiplánicas y su relación con la actividad trófica. Actas del II. Simposio Internacional de Estudios Altiplánicos, Arica, Chile: 135-142.

LOWENGREEN, C., F. OJEDA & V. MONTECINO, 1994. Spectral composition of the aquatic lightfield of the Lakes Rinihue, Todos los Santos, Laguna Negra and El Yeso Reservoir. Arch. Hydrobiol., **129**: 497-509.

LUEBERT, F. & P. PLISCOFF, 2006. Sinopsis bioclimática y vegetacional de Chile: 1-316. (Editorial Universitaria, Santiago de Chile).

MALY, E. J., 1996. A review of relationship among centropagid copepod genera and some species found in Australasia. Crustaceana, **69** (6): 727-733.

MALY, E. J., S. A. HALSE & M. P. MALY, 1997. Distribution of *Boeckella*, *Calamoecia* and *Hemiboeckella* (Copepoda, Calanoida) in Western Australia. Mar. freshwat. Res., **48**: 615-621.

MARINONE, M. C., S. MENU-MARQUE, D. AÑÓN SUÁREZ, M. C. DIEGUEZ, A. P. PÉREZ, P. DE LOS RÍOS, D. SOTO & H. E. ZAGARESE, 2006. UV radiation as a potential driving force for zooplankton community structure in Patagonian lakes. Photochem. Photobiol., **82**: 962-971.

MEDINA, M. & P. PAEZ, 1999. *Artemia* spp., en Laguna del Cisne (Crustacea, Branchiopoda, Anostraca, Artemiidae): registro más austral del género. Noticiario mens. Mus. nac. Hist. nat. Chile, **337**: 17-18.

MENU-MARQUE, S. & E. G. BALSEIRO, 2000. *Boeckella antiqua* n. sp. (Copepoda, Calanoida) from Patagonia. Hydrobiologia, **429**: 1-7.

MENU-MARQUE, S., J. J. MORRONE & C. LOCASCIO DE MITROVICH, 2000. Distributional patterns of the South American species of *Boeckella* (Copepoda, Centropagidae): a track analysis. Journ. Crust. Biol., **20**: 262-272.

MODENUTTI, B. E., E. G. BALSEIRO, C. P. QUEIMALIÑOS, D. A. AÑÓN SUÁREZ, M. C. DIEGUEZ & R. J. ALBARIÑO, 1998. Structure and dynamics of food webs in Andean lakes. Lake Reservoir Management, **3**: 179-189.

MODENUTTI, B. E. & G. L. PÉREZ, 2003. Planktonic ciliates from an oligotrophic South Andean lake, Morenito Lake (Patagonia, Argentina). Brazilian Journ. Biol., **61**: 389-395.

MONTECINO, V., 1991. Primary productivity in South American temperate lakes and reservoirs. Rev. Chilena Hist. nat., **64**: 55-567.

MONTECINO, V., J. P. OYANEDEL, I. VILA & L. ZÚÑIGA, in press. Ecosystem components of natural and man made lakes: honouring Bernard Dussart concerning the knowledge of crustacean zooplankton. Crustaceana Monographs.

MORRIS, D. P., H. E. ZAGARESE, C. E. WILLIAMSON, E. G. BALSEIRO, B. R. HARGREAVES, B. E. MODENUTTI, R. E. MOELLER & C. P. QUEIMALIÑOS, 1995. The attenuation of solar UV radiation in lakes and the role of dissolved organic carbon. Limnol. Oceanog., **40**: 1381-1391.

MRÁZEK, A., 1901. Süsswasser-Copepoden. Ergebn. Hamburger Magalhanischer Sammelreise, **2**: 1-29.

MÜHLHAUSER, H., 1997 Significado de la estructura y funcionamiento de ecosistemas acuáticos y zonas ecotonales altiplánicos para su evaluación, gestión ambiental y conservación. In: Actas del II. Simposio Internacional de Estudios Altiplánicos, Arica, Chile: 127-134.

MÜHLHAUSER, H. & I. VILA, 1987. Eutrofización, impacto en un ecosistema acuático montañoso. Archivos Biol. Med. exp., **20**: 117-124.

MUÑOZ, E., G. MENDOZA & C. VALDOVINOS, 2001. Evaluación rápida de la biodiversidad en cinco sistemas lénticos de Chile central: macroinvertebrados bentónicos. Gayana, Zoología, **65**: 173-180.

NIEMEYER, H. & P. CERECEDA, 1984. Geografía de Chile. Hidrografía: 1-320. (Instituto Geográfico Militar, Santiago de Chile).

OLIVIER, S. R., 1962. Los cladóceros argentinos con claves de las especies, notas biológicas y distribución geográfica. Rev. Mus. La Plata, (n. ser.) (Zoología) **7**: 173-269.

PAGGI, J. C., 1994. Ecología alimentaria de *Branchinecta gaini* (Crustacea, Anostraca) en lagunas de la península Potter, Isla 25 de Mayo, Shetland del Sur, Antártida. Tankay, **1**: 111-112.

— —, 1996a. *Daphnia* (*Ctenodaphnia*) *menucoensis* (Anomopoda; Daphniidae) a new species from athalassic saline waters in Argentina. Hydrobiologia, **319**: 137-147.

— —, 1996b. Feeding ecology of *Branchinecta gaini* (Crustacea: Anostraca) in ponds of South Shetland Islands, Antarctica. Polar Biol., **16**: 13-18.

— —, 1999. Status and phylogenetic relationships of *Daphnia sarsi* Daday, 1902 (Crustacea, Anomopoda). Hydrobiologia, **403**: 27-38.

PEDROZO, F., S. CHILLRUD, P. TEMPORETTI & M. DÍAZ, 1993. Chemical composition and nutrient limitation in river and lakes of northern Patagonian Andes (39.5-42°S, 71°W) (Rep. Argentina). Verh. intern. Verein. angew. Limnol., **25**: 207-214.

PEZZANI-HERNÁNDEZ, S., 1970. Cladóceros del embalse del río Yeso (Crustacea, Cladocera). Noticiario mens. Mus. nac. Hist. nat. Santiago de Chile, **16** (168): 3-9.

PILATI, A., 1997. Copépodos calanoideos de la provincia de la Pampa (Argentina). Revista Fac. agron. Univ. nac. La Plata, **9**: 57-66.

— —, 1999. Copépodos ciclopoideos en la provincia de la Pampa (Argentina). Revista Fac. agron. Univ. nac. La Plata., **10**: 29-44.

PILATI, A. & S. MENU-MARQUE, 2003. Morphological comparison of *Mesocyclops araucanus* Campos et al., 1974, and *M. longisetus* Thiébaud, 1912, and first description of their males. Beaufortia, **52**: 45-62.

PINDER, A. M., S. A. HALSE, J. M. MCRAE & R. S. SHIELL, 2005. Occurrence of aquatic invertebrates of the wheatbelt region of Western Australia in relation to salinity. Hydrobiologia, **543**: 1-24.

PINTO-COELHO, R., B. PINEL-ALLOUL, G. METHOT & K. E. HAVENS, 2005. Crustacean species richness in lakes and reservoirs of temperate and tropical regions: variations with trophic status. Canadian Journ. Fish. aquat. Sci., **62**: 348-361.

POCIECHA, A. & H. J. DUMONT, 2008. Life cycle of *Boeckella poppei* Mrázek and *Branchinecta gaini* Daday (King George Island, South Shetlands). Polar Biol., **31**: 245-248.

PUGH, P., H. DARTNALL & S. MCINNES, 2002. The nonmarine Crustacea of Antarctica and the islands of the Southern Ocean: biodiversity and biogeography. Journ. nat. Hist., London, **36**: 1047-1103.

QUIROS, R. & E. DRAGO, 1999. The environmental state of Argentinean lakes: an overview. Lake Reservoir Management, **4**: 55-64.

RAMOS, R., C. TRAPO, F. FLORES, A. BRIGNARDELLO, O. SIEBECK & L. ZÚÑIGA, 1998. Temporal succession of planktonic crustaceans in a small temperate lake (El Plateado, Valparaíso, Chile). Verh. intern. Verein. angew. Limnol., **26**: 1997-2000.

RAMOS-JILIBERTO, R. & L. R. ZÚÑIGA, 2001. Depth-selection patterns and diel vertical migration of *Daphnia ambigua* (Crustacea, Cladocera) in Lake El Plateado. Rev. Chilena Hist. nat., **74**: 573-585.

RAU, J., A. GRANTZ, L. MONTENEGRO, A. APARICIO, P. VARGAS-ALMONACID, M. E. CASANUEVA, J. STUARDO & J. E. CRESPO, 2006. Diversidad de árboles, micromoluscos y aves en hábitats fragmentados del centro-sur de Chile. In: A. A. GREZ, J. A. SIMONETTI & R. O. BUSTAMENTE (eds.), Biodiversidad y patrones en ambientes fragmentados de Chile: patrones y procesos a diferentes escalas: 143-148. (Editorial Universitaria, Santiago de Chile).

RAUTIO, M. & A. KORKHOLA, 2002a. UV-induced pigmentation in Subarctic *Daphnia*. Limnol. Oceanogr., **47**: 295-299.

— — & — —, 2002b. Impacts of UV radiation on the survival of some key Subarctic crustaceans. Polar Biol., **25**: 460-468.

RECHE, I., M. L. PACE & J. J. COLE, 1998. Interactions of photobleaching and inorganic nutrients in determining bacterial growth on colored dissolved organic carbon. Microb. Ecol., **36**: 270-280.

REID, J., 1985. Chave de identifição e lista de referencias para as species continentais sudamericanas de vida livre da ordem Cyclopoida (Crustacea, Copepoda). Bolm. Zool. Univ. São Paulo, **9**: 17-143.

REISSIG, M., B. MODENUTTI, E. BALSEIRO & C. QUEIMALIÑOS, 2004. The role of the predaceous copepod *Parabroteas sarsi* in the pelagic food web of a large deep Andean lake. Hydrobiologia, **524**: 67-77.

RHODE, S. C., M. PAWLOWSKI & R. TOLLRIAN, 2001. The impact of ultraviolet radiation on the vertical distribution of zooplankton of the genus *Daphnia*. Nature, London, **412**: 69-72.

RICHASER, F., H. ALONSO & C. SALAZAR, 1999. Geoquímica de aguas en cuencas cerradas: I, II y III regiones, Chile: 221-230. (Technical Report of the Dirección General de Aguas, Chile, Universidad Católica del Norte, Chile, and the Institut de Recherche pour le Développement, France). (Chilean Second District.)

ROCCO, V. E., O. OPEZZO, R. PIZARRO, R. SOMMARUGA, M. FERRARO & H. E. ZAGARESE, 2002. Ultraviolet damage and counteracting mechanisms in the freshwater copepod *Boeckella poppei* in the Antarctic Peninsula. Limnol. Oceanogr., **47**: 829-836.

ROGERS, C., P. DE LOS RÍOS & O. ZÚÑIGA, 2008. Fairy shrimp (Branchiopoda) of Chile. Journ. Crust. Biol., **28**: 543-550.

ROKNEDDINE, A., 2004a. Effets de la salinité et de la température sur la croissance et la reproduction de *Moina salina* Daday, 1888 (Branchiopoda, Moinidae). Crustaceana, **77** (7): 805-824.

— —, 2004b. Influence de la salinité et de la temperature sur la reproduction d'*Arctodiaptomus salinus* (Daday, 1885) (Copepoda, Calanoida), du marais temporaire salé "La Sebkha Zima" (Maroc). Crustaceana, **77** (8): 923-940.

— —, 2005. Influence de la salinité et de la temperature sur la croissance d'*Arctodiaptomus salinus* (Daday, 1885) (Copepoda, Calanoida), du marais temporaire salé "La Sebkha Zima", Maroc. Crustaceana, **77** (9): 1025-1044.

RUIZ, R. & N. BAHAMONDE, 1989. Cladóceros y copépodos límnicos en Chile y su distribución geográfica. Lista sistemática. Publicaciones Ocasional, Mus. nac. Hist. nat., Santiago de Chile, **45**: 1-48.

— — & — —, 2003. Distribución estacional de cladóceros y copépodos en el lago Rapel, Chile central. Publicaciones Ocasional, Mus. nac. Hist. nat., Santiago de Chile, **58**: 5-58.

SAKAMOTO, M. & T. HANAZATO, 2008. Antennule shape and body size of *Bosmina*: key factors that determine its vulnerability to predaceous copepods. Limnology, **9**: 27-34.

SCHLATTER, R. & W. SIELFIELD, 2006. Avifauna y mamíferos acuáticos de humedales en Chile. In: I. VILA, A. VELOSO, R. SCHLATTER & C. RAMIREZ (eds.), Macrófitas y vertebrados de los sistemas límnicos de Chile: 141-186. (Editorial Universitaria, Santiago de Chile).

SCHMID-ARAYA, J. M. & L. R. ZUÑIGA, 1992. Zooplankton community structure in two Chilean reservoirs. Arch. Hydrobiol., **123**: 305-335.

SILVA, V. M., 1998. Diversity and distribution of the free-living freshwater Cyclopoida (Copepoda: Crustacea) in Neotropics. Brazilian Journ. Biol., (Suppl.) **68**: 1099-1106.

SORGELOOS, P., P. LAVENS, P. LEGER, W. TACKAERT & D. VERSICHELE, 1986. Manual for the culture and use of brine shrimp *Artemia* in aquaculture. Prepared for the Belgian Administration for Development Cooperation: 1-391. (Food and Agricultural Organization of the United Nations; State University of Ghent, Belgium, Faculty of Agriculture, Ghent).

SOTO, D., 1990. Relationship between zooplankton biomass and Chilean flamingo population in south Chile Patagonic lagoons. In: Actas del I. Taller Internacional de Especialistas en Flamencos Sudamericanos: 90-115. (Corporación Nacional Forestal de Chile and Zoological Society of New York).

— —, 2002. Oligotrophic patterns in southern Chile lakes: the relevance of nutrients and mixing depth. Rev. Chilena Hist. nat., **75**: 377-393.

SOTO, D. & H. CAMPOS, 1995. Los lagos oligotróficos del bosque templado húmero del sur de Chile. In: J. ARMESTO, M. KHALIN & M. VILLAGRÁN (eds.), Ecología del Bosque Chileno: 134-148. (Editorial Universitaria, Santiago de Chile). [In Spanish with English summary.]

SOTO, D., H. CAMPOS, W. STEFFEN, O. PARRA & L. ZÚÑIGA, 1994. The Torres del Paine lake district (Chilean Patagonia): a case of potentially N-limited lakes and ponds. Arch. Hydrobiol., **99**: 181-197.

SOTO, D. & P. DE LOS RÍOS, 2006. Trophic status and conductivity as regulators of daphnids dominance and zooplankton assemblages in lakes and ponds of Torres del Paine National Park. Biologia, Bratislava, **61**: 541-546.

SOTO, D. & J. G. STOCKNER, 1996. The temperate rain forest lakes of Chile and Canada: comparative ecology and sensitivity to anthropogenic change. In: R. LAWFORD, P. ALABACK & E. FUENTES (eds.), High latitude rain forest of the west coast of the Americas. Climate, hydrology, ecology and conservation: 266-280. (Springer, New York).

SOTO, D. & L. R. ZÚÑIGA, 1991. Zooplankton assemblages of Chilean temperate lakes: a comparison with North American counterparts. Rev. Chilena Hist. nat., **64**: 569-581.

SPENCER, M., L. BLAUSTEIN, S. S. SCHWARTZ & R. COHEN, 1999. Species richness and the proportion of predatory animal species in temporary freshwater pools: relationships with habitat size and permanence. Ecol. Lett., **2**: 157-166.

SPENCER, M., S. S. SCHWARTZ & L. BLAUSTEIN, 2002. Are there fine-scale spatial patterns in community similarity among temporary freshwater pools? Glob. Ecology Biogeogr., **11**: 71-78.

STEINHART, G. S., G. E. LIKENS & D. SOTO, 1999. Nutrient limitation in Lago Chaiquenes (Parque Nacional Alerce Andino, Chile): evidence from nutrient experiments and physiological assays. Rev. Chilena Hist. nat., **72**: 559-560.

— —, — — & — —, 2002. Physiological indicators of nutrient deficiency in phytoplankton of southern Chilean lakes. Hydrobiologia, **489**: 21-27.

STERNER, R. W., J. J. ELSER, E. J. FEE, S. J. GUILDFORD & T. H. CHRZANOWSKI, 1997. The light : nutrient ratio in lakes: the balance of energy and materials affects ecosystems structure and process. American Nat., **150**: 663-684.

STERNER, R. W. & D. O. HESSEN, 1994. Algal nutrient limitation and nutrient of aquatic herbivores. Annu. Rev. Ecol. Syst., **25**: 1-29.

STORZ, U. C. & R. J. PAUL, 1998. Phototaxis in water fleas (*Daphnia magna*) is differently influenced by visible and UV light. Journ. comp. Physiol., (A) **183**: 709-717.

TARTAROTTI, B., G. BAFFICO, P. TEMPORETTI & H. ZAGARESE, 2004. Mycosporine-like amino acids in planktonic organisms living under different UV exposure conditions in Patagonian lakes. Journ. Plankt. Res., **26**: 753-762.

THOMASSON, K., 1963. Araucanian lakes. Acta Phytogeographica Suecica, **47**: 1-139.

TRIANTAPHYLLIDIS, G. V., T. J. ABATZOPOULOS & P. SORGELOOS, 1998. A review of the biogeography of the genus *Artemia* (Crustacea, Anostraca). Journ. Biogeogr., **25**: 213-226.

TROCHINE, C., E. BALSEIRO & B. MODENUTTI, 2008. Zooplankton of fishless ponds of northern Patagonia: insights into predation effects of *Mesostoma ehrenbergii*. Intern. Revue ges. Hydrobiol., **93**: 312-327.

VALERO-GARCÉS, B. L., M. GROSJEAN, A. SCHWALB, M. GEYH, B. MESSERLI & K. KELTS, 1996. Limnogeology of Laguna Miscanti: evidence of moisture changes in the Atacama Altiplano (northern Chile). Journ. Paleolimnol., **16**: 1-21.

VARESHI, E. & D. WUBBENS, 2001. Vertical migration of *Daphnia pulex* in response to UV radiation. Verh. intern. Verein. angew. Limnol., **27**: 3349-3353.

VEGA, M., 1996. Morphology and defensive structures in the predator-prey interaction: an experimental study of *Parabroteas sarsi* (Copepoda, Calanoida) with different cladoceran prey. Hydrobiologia, **299**: 139-145.

— —, 1997. The functional response of copepodid stages to adult of *Parabroteas sarsi* (Copepoda, Calanoida). Intern. Revue ges. Hydrobiol., **154**: 647-663.

— —, 1998. Impacts of *Parabroteas sarsi* (Copepoda, Calanoida) predation on planktonic cladocerans in a pond of the southern Andes. Journ. freshwat. Ecol., **13**: 383-389.

— —, 1999. Life-stage differences in the diet of *Parabroteas sarsi* (Daday) (Copepoda, Calanoida): a field study. Limnológica, **29**: 186-190.

VIGNATTI, A., S. ECHANIZ & M. C. MARTÍN, 2007. El zooplankton de tres lagos someros de diferente salinidad y estado trófico en la región semiárida pampeana (Argentina). Gayana, Zoología, **71**: 34-48.

VILA, I., R. PARDO, B. DYER & E. HABIT, 2006. Peces limnicos: diversidad, origen y estado de conservación. In: I. VILA, A. VELOSO, R. SCHLATTER & C. RAMIREZ (eds.), Macrófitas y vertebrados de los sistemas límnicos de Chile: 73-102. (Editorial Universitaria, Santiago de Chile).

VILLAFAÑE, V. E., E. S. BARBIERI & W. E. HELBLING, 2004. Annual patterns of ultraviolet radiation effects on temperate marine phytoplankton of Patagonia Argentina. Journ. Plankt. Res., **26**: 167-174.

VILLAFAÑE, V. E., E. W. HELBLING & H. E. ZAGARESE, 2001. Solar ultraviolet radiation and its impact on aquatic ecosystems of Patagonia, South America. Ambio, **30**: 112-117.

VILLALOBOS, L., 1994. Distribution of *Daphnia* in high mountain and temperate lakes of South America. Verh. intern. Verein. angew. Limnol., **25**: 2400-2404.

— —, 1999. Determinación de capacidad de carga y balance de fósforo y nitrógeno de los lagos Riesco, Los Palos, y Laguna Escondida en la XI región. (Technical Report Fisheries Research Foundation-Chile, FIP-IT/97-39.)

— —, 2002. Comparison of the filtration structures in South American *Daphnia*. Arch. Hydrobiol., **154**: 647-663.

Villalobos, L. & W. Geller, 1997. Setular bosses: report of a new ultrafine structure on the filter appendages of *Daphnia*. Arch. Hydrobiol., **140**: 565-575.

Villalobos, L., O. Parra, M. Grandjean, E. Jaque, S. Wölfl & H. Campos, 2003. River basin and limnological study in five humic lakes of the Chiloé Island. Rev. Chilena Hist. nat., **76**: 10-15.

Villalobos, L., S. Wölfl, O. Parra & H. Campos, 2003. Lake Chapo: a baseline of a deep, oligotrophic North Patagonian lake prior to its use for hydroelectricity generation: II. Biological properties. Hydrobiologia, **510**: 225-237.

Wetzlar, H., 1979. Beiträge zur Biologie und Bewirtschaftung von Forellen (*Salmo gairdneri* und *Salmo trutta*) in Chile: 1-264. (Ph.D. Thesis, University of Freiburg, Freiburg).

Williams, W. D., 1998. Salinity as a determinant of the structure of biological communities in salt lakes. Hydrobiologia, **381**: 191-201.

Williams, W. D., T. R. Carrick, I. A. E. Bayly, J. Green & D. B. Herbst, 1995. Invertebrates in salt lakes of the Bolivian Altiplano. Intern. Journ. Salt Lake Res., **4**: 65-77.

Winder, M., 2001. Zooplankton ecology in high mountain lakes: 1-157. (Doctoral Thesis submitted to the Swiss Federal Institute of Technology, Zürich).

Wölfl, S., 1996. Untersuchungen zur Zooplanktonstruktur einschliesslich der mikrobiellen Gruppen unter besonderer Berücksichtigung der mixotrophen Ciliaten in zwei südchilenischen Andenfußseen: 1-242. (Doctoral Thesis, University of Konstanz).

— —, 2007. The distribution of large mixotrophic ciliates (*Stentor*) in deep North Patagonian lakes (Chile): first results. Limnologica, **37**: 28-36.

Wölfl, S. & W. Geller, 2002. *Chlorella*-bearing ciliates dominant in an oligotrophic North Patagonian lake (Pirihueico Lake, Chile): abundance, biomass and symbiotic photosynthesis. Freshwat. Biol., **47**: 231-242.

Wölfl, S., L. Villalobos & O. Parra, 2003. Trophic parameters and method validation in a Lake Riñihue (North Patagonia, Chile) from 1978 to 1997. Rev. Chilena Hist. nat., **76**: 459-474.

Zagarese, H. E., B. Tartarotti, W. Cravero & P. González, 1998. UV damage in shallow lakes: the implications of water mixing. Journ. Plankt. Res., **20**: 1423-1433.

Zeller, M., R. Jimenez-Melero & B. Santer, 2004. Diapause in the calanoid freshwater copepod *Eudiaptomus graciloides*. Journ. Plankt. Res., **26**: 1379-1388.

Zúñiga, L., 1975. Sobre *Diaptomus diabolicus* Brehm (Crustacea: Copepoda, Calanoida). Noticiario mens. Mus. nac. Hist. nat., Santiago de Chile, **19** (228): 3-9.

Zúñiga, L. R. & J. M. Araya, 1982. Estructura y distribución especial del zooplankton del embalse Rapel. Anales Mus. Hist. nat. Valparaíso, **15**: 45-57.

Zúñiga, L., V. Campos, H. Pinochet & B. Prado, 1991. A limnological reconnaissance of Lake Tebenquiche, Salar de Atacama, Chile. Hydrobiologia, **210**: 19-25.

Zúñiga, L. R. & P. Domínguez, 1977. Observaciones sobre el zooplancton de lagos chilenos. Anales Mus. Hist. nat. Valparaíso, **10**: 107-120.

— — & — —, 1978. Entomostracos planctónicos del lago Riñihue (Valdivia, Chile): distribución temporal de la taxocenosis. Anales Mus. Hist. nat. Valparaíso, **11**: 89-95.

Zúñiga, O., R. Wilson, F. Amat & F. Hontoria, 1999. Distribution and characterization of Chilean populations of the brine shrimp *Artemia* (Crustacea, Branchiopoda, Anostraca). Intern. Journ. Salt Lake Res., **8**: 23-40.

Zúñiga, O., R. Wilson, R. Ramos, E. Retamales & L. Tapia, 1994. Ecología de *Artemia franciscana* en la laguna Cejas, Salar de Atacama, Chile. Estudios Oceanológicos, **13**: 71-84.

TAXONOMIC INDEX